Technical Drawing for CSE

Malcolm McAlpine and
Alexander B Goad

Longman

Longman
1724-1974

Longman Group Limited
London
Associated companies, branches and representatives throughout the world.

© Longman Group Limited 1974
All rights reserved. No part of this publication may be reproduced, stored in a retreival system or transmitted in any form or by any means – electronic, mechanical, photocopying, recording or otherwise – without the prior permission of the copyright owner.

First published 1974

ISBN O 582 220076

Printed photolitho in Great Britain by
J. W. Arrowsmith Ltd., Bristol

Acknowledgements

We are grateful to the following Examining Boards for permission to reproduce questions from past examination papers:
East Anglian Examinations Board; East Midland Regional Examinations Board; Middlesex Regional Examining Board; North Western Secondary School Examinations Board; The South-East Regional Examinations Board and The West Yorkshire and Lindsey Regional Examining Board.

Preface

TECHNICAL DRAWING FOR C S E
Although this book has been designed primarily for those pupils who will follow a course leading to the C S E, it should prove useful in the early stages of other courses. The intention is that a pupil can work at his own speed with the minimum of help and direction. Therefore, to cover a wide ability range, the processes have been broken down into very small steps, on the assumption that it is easier for some pupils to telescope stages together than it is for others to break down large stages into their components.

While all the examined topics likely to be encountered in the C S E examination have been covered, with plenty of examples for pupils to work, the nature of the approach adopted in this book makes it impossible, within the compass of two hundred pages, to deal with them exhaustively. Once the pupil has used the book to master a particular process he can be presented with more conventional material.

As far as is possible in geometrical drawing the recommendation of B.S.308 has been followed and metric units are used throughout.

Contents

Angles	2
Triangles	14
Division of lines	20
Scales	24
Diagonal Scales	26
Regular polygons	30
Hexagon	30
Octagon	36
Heptagon	38
Pentagon	40
Equivalent area	42
Similar figures	48
The circle	54
Tangents	56
Rounding off corners	64
Circle to touch a given circle and straight line	72
Circles in contact	76
Common tangents	88
Circles and triangles	92
Loci	94
Mechanisms	94
Involute	100
Cycloid	102
Parabola	104
The helix	106
The ellipse	108
Locus from given definition	112
Orthographic projection	114
First angle	114
Third angle	120
Sections	124
Auxiliary views	126
Isometric projection	136
Oblique projection	178
Developments	184
Interpenetration	192
True length	204
Examples	212

ANGLES

If a line AO turns about O to a new position A_1O, as shown in Panel No. 1 then it has turned through an angle. This movement is measured in degrees.

If the line continues turning until it has reached its original position it will have turned through a complete circle of 360°. In turning through a quarter of a circle it would turn through 90° and through a semi-circle 180°.

Panel No. 3 shows a semi-circular protractor which is used for measuring angles. Notice that it is numbered 0 to 180° both clockwise and counter-clockwise. In the illustration it is being used to draw an angle AOB of 30°.

Panel No. 4 shows the protractor being used to measure an angle AOB of 25°.

Panel No. 5 shows the protractor being used to draw an angle COB of 155°.

To draw a quadrilateral with unequal sides and angles. Draw this exercise following, stage by stage, the instructions and the drawings in the panels.

Panel No:

6 Draw line AB 45 mm long.

7 Draw a line at A making an angle of 110° with AB and another line at B making an angle of 115°. From B measure 75 mm to fix point C.

8 Draw a line from C making an angle of 50° with line BC.

9 Continue the four sides of the quadrilateral as shown and measure each of the four external angles indicated. Add these four angles together — what is their total?

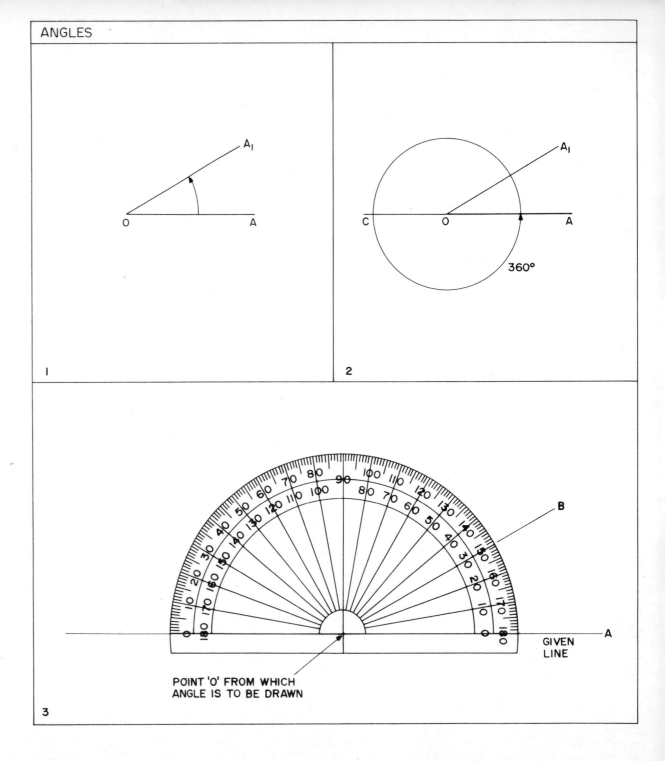

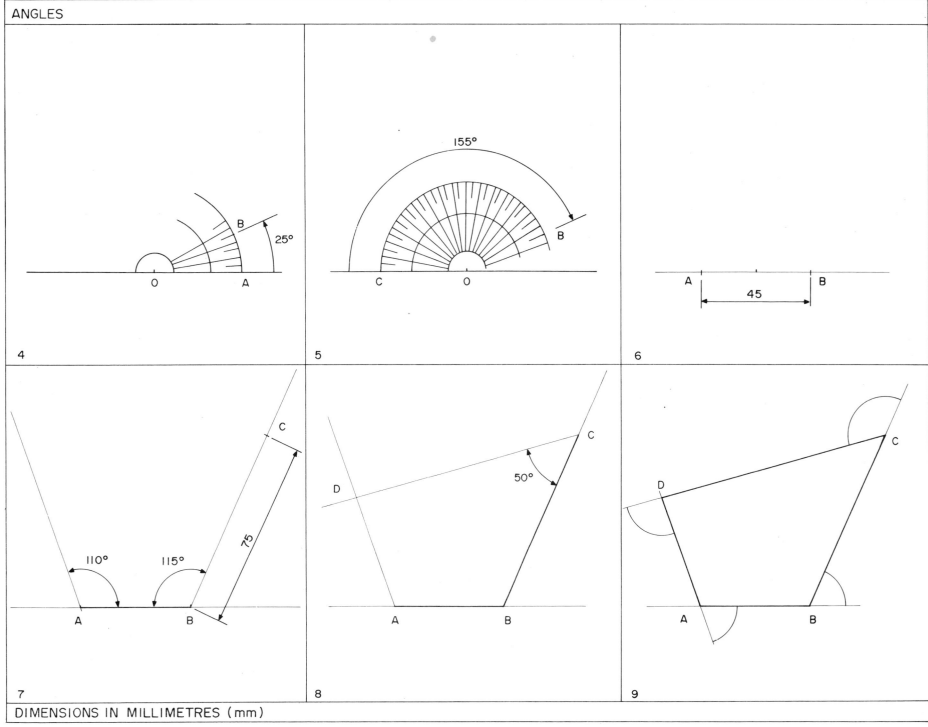

Some angles can be drawn using a tee square and set squares. You will normally have two set squares, one with an angle of 90°, and two angles of 45°, the other with one angle of 90°, one of 60° and one of 30°.

To set out the triangle in Panel No. 2.
Draw this exercise following, stage by stage, the instructions and the drawings in the panels.

Panel No:
1. This shows some of the angles which can be drawn with the two set squares and the tee square.
2. This shows a triangle ABC in which the length of the base and the two base angles are given.
3. Draw a faint horizontal line using your tee square.
4. Mark off point A and C 100 mm apart.
5. Line in the base AC.
6. With your tee square and 45° set square, draw a faint line from A at an angle of 45° with AC.
7. With your tee square and 60°/30° set square draw a faint line from C at 60° to AC and line in the required portion CB.
8. Line in AB.

Exercise:
9. Set out the triangle to the dimensions shown and measure the length of side AB.

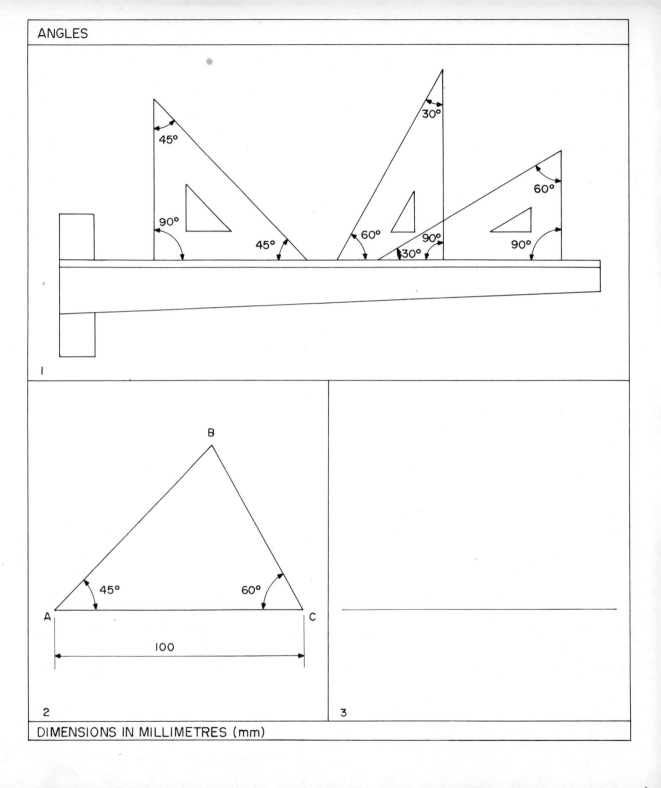

DIMENSIONS IN MILLIMETRES (mm)

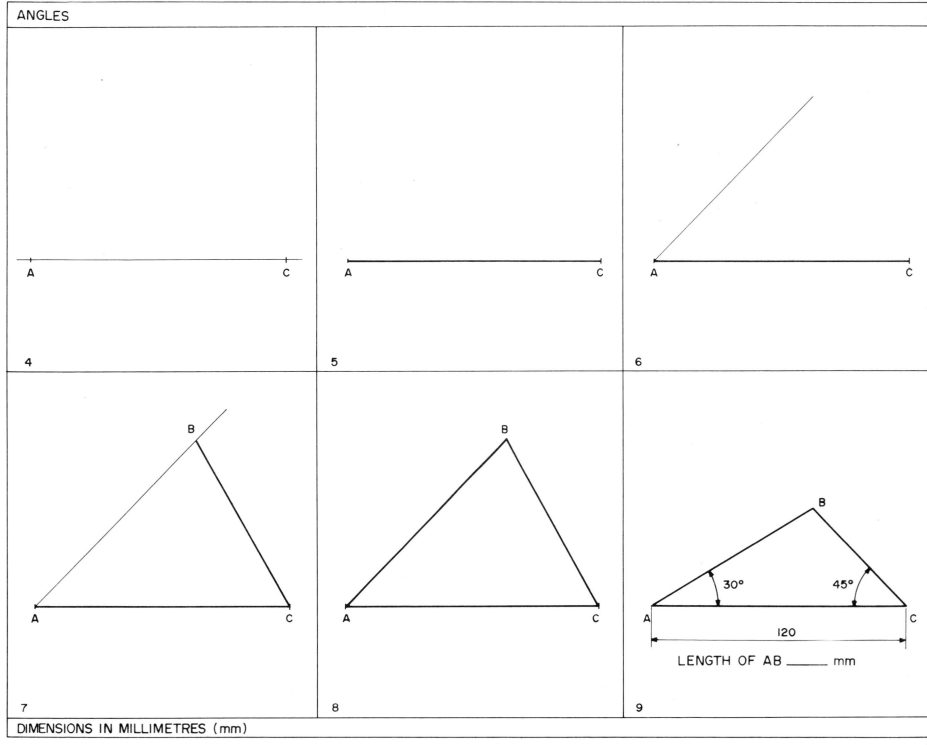

To set out the quadrilateral (a four sided figure) in Panel No. 2.

Draw this exercise following, stage by stage, the instructions and the drawings in the panels.

Panel No:

1. This shows you how, by combining your two set squares, you can draw angles of 75° and 105°. Notice these add up to 180° which is the angle of a straight line.

2. This shows a quadrilateral with sufficient details to enable it to be drawn.

3. With your tee square draw a faint horizontal line and mark on it points A and D 150 mm apart. Line in.

4. At D draw a line at 90° to AD and mark point C 90 mm along it.

5. Using both set squares draw a line from A at 75° to AD. From C draw a line at 135° to DC to fix point B.
(Note: 135° = 90° + 45°)

6. Line in the quadrilateral.

Exercises:

7. Draw the given quadrilateral – full size.

8. Draw the given pentagon full size. Work out the size of angle θ before starting.

9. Draw the given pentagon – full size.

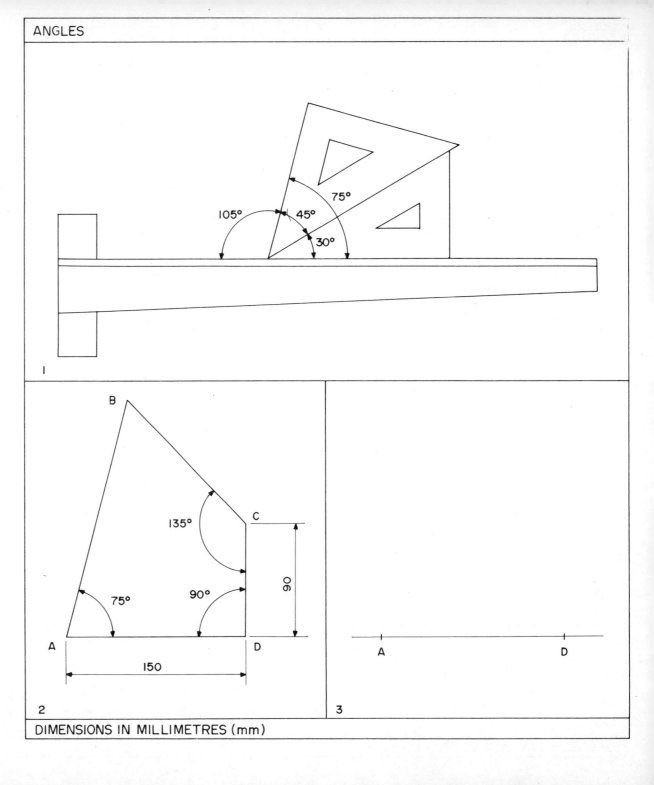

DIMENSIONS IN MILLIMETRES (mm)

ANGLES

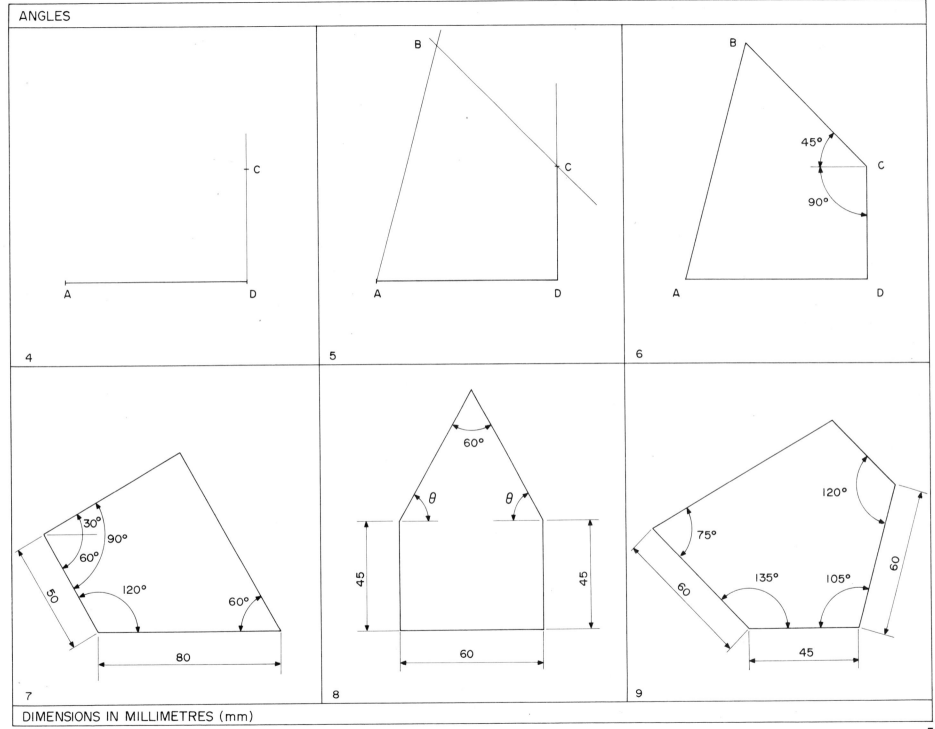

DIMENSIONS IN MILLIMETRES (mm)

It is sometimes necessary to draw a line at right angles to another and passing through its centre. This line is called the perpendicular bisector.

To bisect a given line AB
Draw this exercise following, stage by stage, the instructions and the drawings in the panels.

Panel No:
1. Draw a line AB 107 mm long.

2. Set your compasses to any length provided it is greater than half AB. With the point of the compasses on A draw arc CD.

3. Without altering your compasses put their point on B and draw an arc to intersect the arc CD.

4. Draw a straight line from C to D. This is the perpendicular bisector of AB — in other words it divides line AB into two equal halves and is at right angles to it.

To copy a given angle BAC.
Draw this exercise following, stage by stage, the instructions and the drawings in the panels.

Panel No:
5. Draw any acute angle BAC. This is the angle to be copied.
   Some distance below it draw a straight line FH.

6. Set your compasses to any measurement and with the point at A draw the arc ED.

7. With the same radius and the point of the compasses at F draw an arc JG.

8. With the point of the compasses at E open them out to draw arc KL.

9. With the point of the compasses at J draw arc $K_1 L_1$ to fix point G.

10. Draw a straight line from F to pass through point G. Angle GFJ should be a copy of angle BAC.

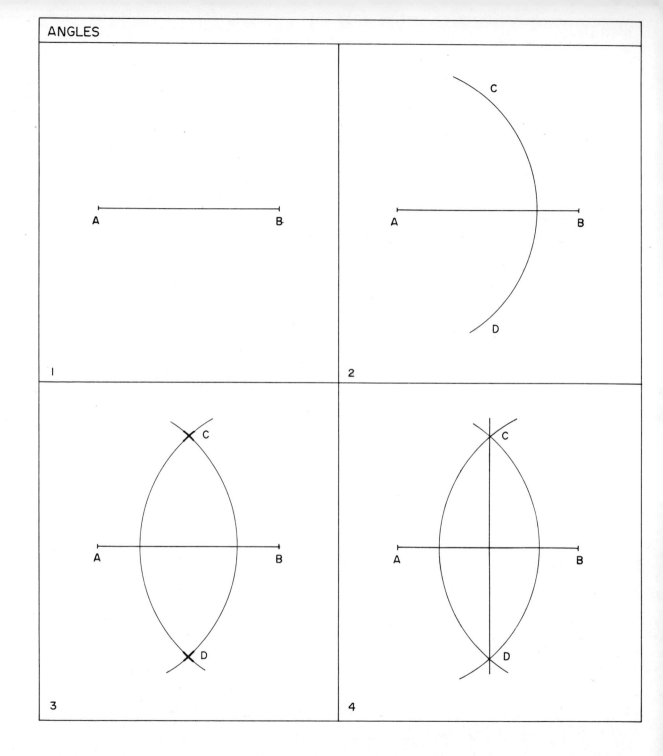

8

ANGLES

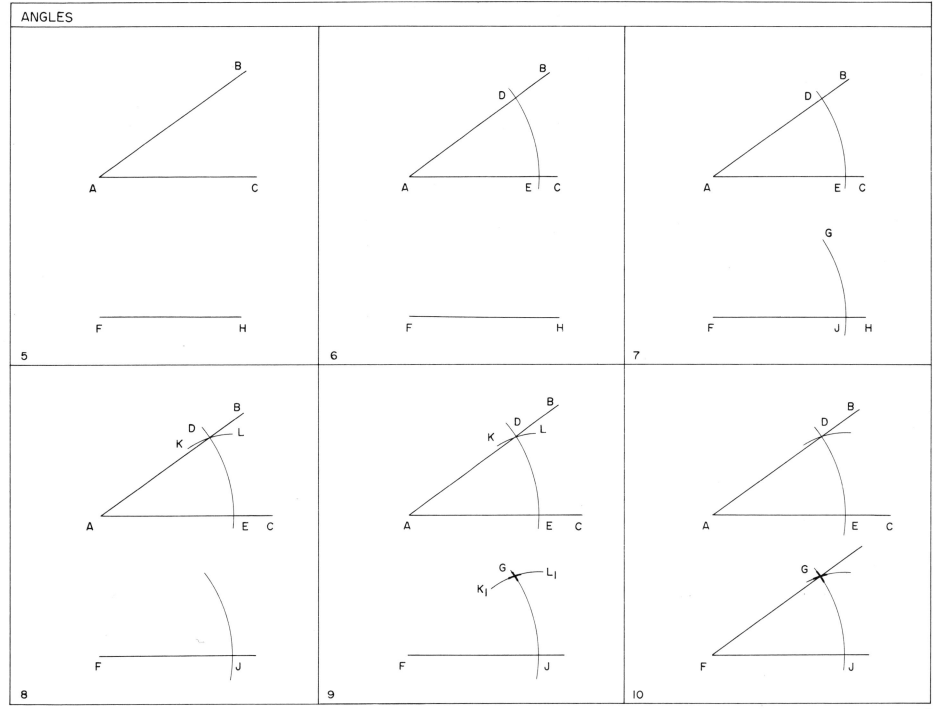

To bisect an angle
Draw this exercise following, stage by stage, the instructions and the drawings in the panels.

Panel No:

1. Draw any angle ABC.

2. From B draw an arc cutting AB and BC at E and D.

3. With centre E draw an arc FG.

4. Using the same radius and centre D draw the arc HJ to intersect FG at K.

5. Join BK the bisector of angle ABC.

To construct a perpendicular to a line from any point.

Panel No:

6. Draw the line AB and mark any point P. Suitable dimensions.
   AB 100 mm. AP 80 mm. BP 70 mm.

7. With centre P draw an arc cutting AB to give points C and D. Choose a radius that results in C and D being a reasonable distance apart.

8. From centre C draw any arc CE.

9. Using the same radius and centre D draw DF intersecting CE at G.

10. Join PG which will be perpendicular to AB.

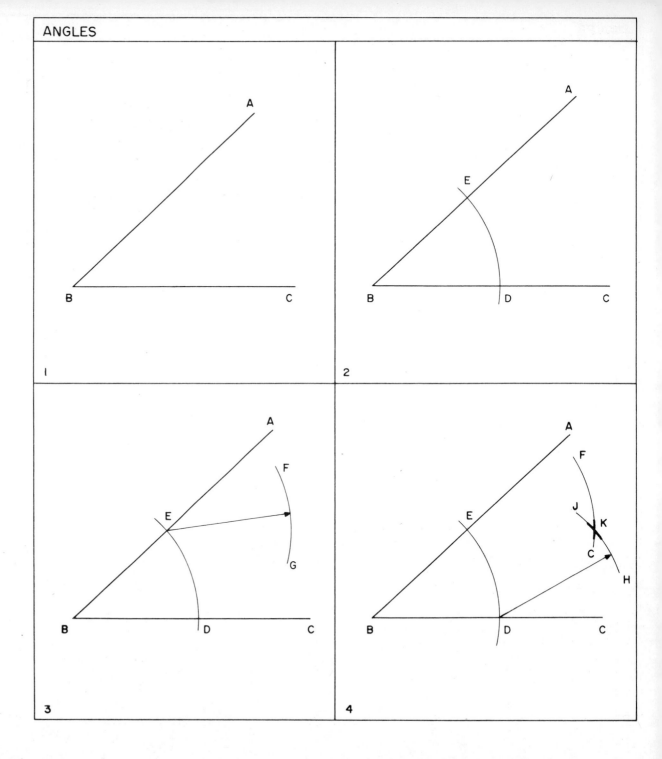

10

ANGLES

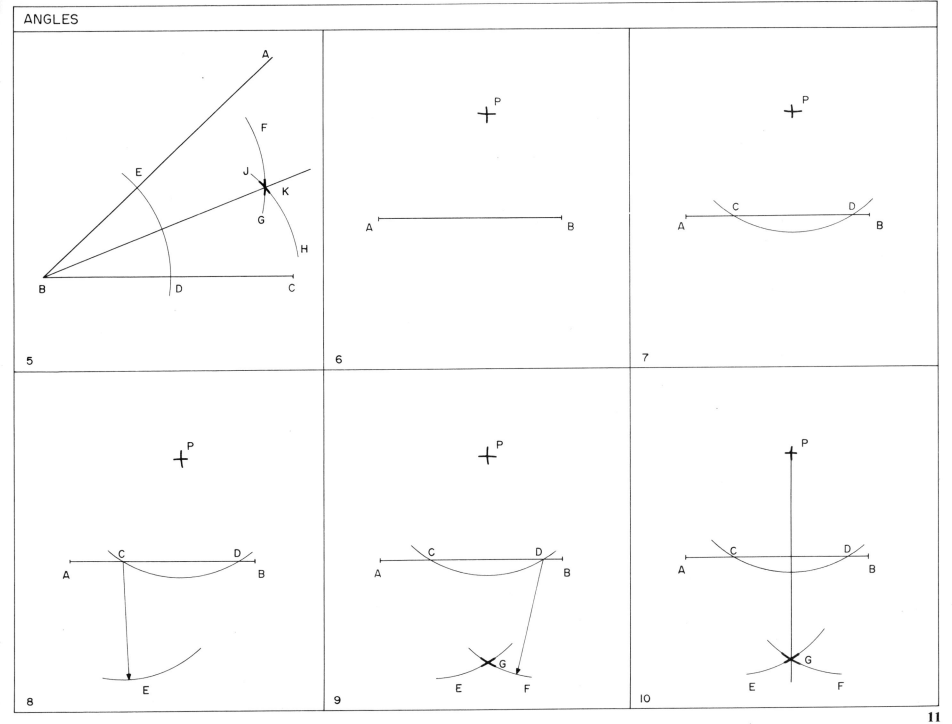

Some angles can be drawn using a straight edge and compasses only and you should be familiar with these constructions.

To construct an angle of 60° using a straight edge and compasses

Draw this exercise following, stage by stage, the instructions and the drawings in the panels.

Panel No:

1. Draw a line AB 100 mm long.
   With your compasses set at 40 mm (or any suitable measurement) draw the arc CD with centre at A.

2. Without altering your compasses draw arc EF with centre at C.
   The intersection of arcs CD and EF fixes point G.

3. Draw a straight line from A through G.

4. You now have an angle of 60° and of course its supplementary angle of 120°.

To construct an angle of 30°.

Panel No:

5. Construct an angle of 60° as you did in the last exercise.

6. Bisect this angle giving two of 30° each.
   (Note: This also enables you to draw the supplementary angle of 150°.)

To construct an angle of 90°.

Panel No:

7. Draw a straight line and mark on it point B.

8. With B as centre draw a semi-circle of any size cutting the line at A and C.

9. Open out your compasses and with the point of the compasses on C, draw arc EF, with the point of the compasses on A draw arc GH.
   The intersection of arcs EF and GH fixes point D.

10. A straight line drawn through B and D forms a right angle with the original line.

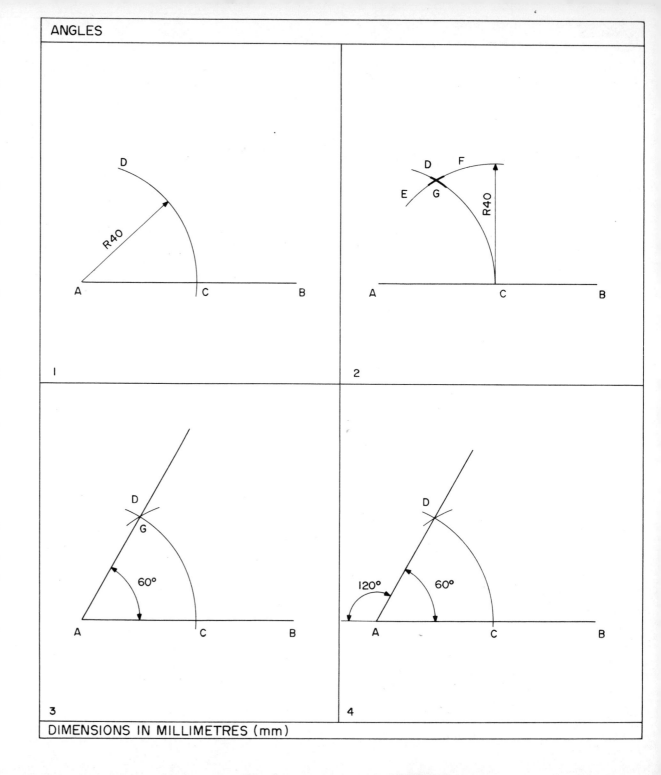

ANGLES

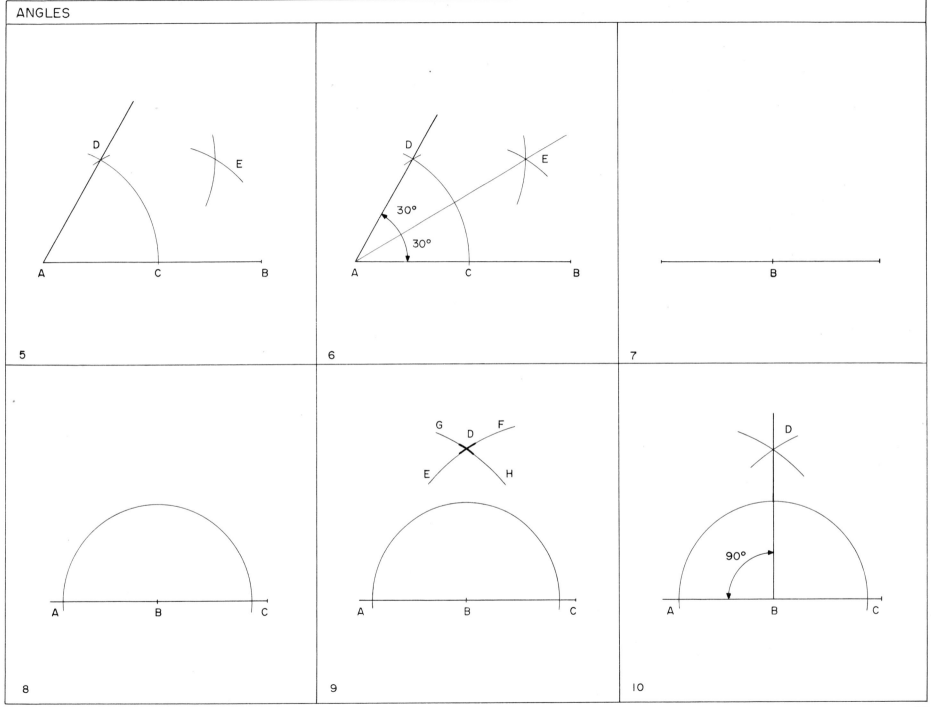

TRIANGLES

A triangle is a plane figure with three straight sides.

Panel No:

1. This shows a SCALENE triangle, that is, one with three sides of different lengths and three angles of different sizes. The 'corners' A B and C are called VERTICES (singular VERTEX).

2. This shows an ISOSCELES triangle, that is, one with two sides equal and two angles equal. The two equal angles are of course the base angles.

3. This shows another ISOSCELES triangle.

4. This shows an EQUILATERAL triangle, that is, one with three equal sides and three equal angles. Each angle will be 60° as the three angles of any triangle when added together come to 180°.

5. This shows an ACUTE ANGLED triangle in which all three angles are less than 90°.

6. This shows an OBTUSE ANGLED triangle in which one angle is greater than 90°.

7. This shows a RIGHT ANGLED triangle in which one angle is 90°. The side opposite the right angle is called the HYPOTENUSE.

8. The ALTITUDE (or VERTICAL HEIGHT) of a triangle is the distance measured from the base (at right angles to it) to the angle opposite the base which is called the APEX.

9. If side AB is regarded as the BASE then C is the APEX and DC is the ALTITUDE.
   If side BC is regarded as the BASE then A is the APEX and FA is the ALTITUDE.
   If side AC is regarded as the BASE then B is the APEX and EB is the ALTITUDE.

10. When an angle such as A is regarded as the APEX then the ALTITUDE has to be measured from a line which is the extension of the BASE.

NB: The perimeter of a triangle is the sum of its three sides.

The area of a triangle is found by using the formula AREA = ½(BASE x HEIGHT)

The HEIGHT is the ALTITUDE or VERTICAL HEIGHT.

The sum of the angles in a triangle is 180°.

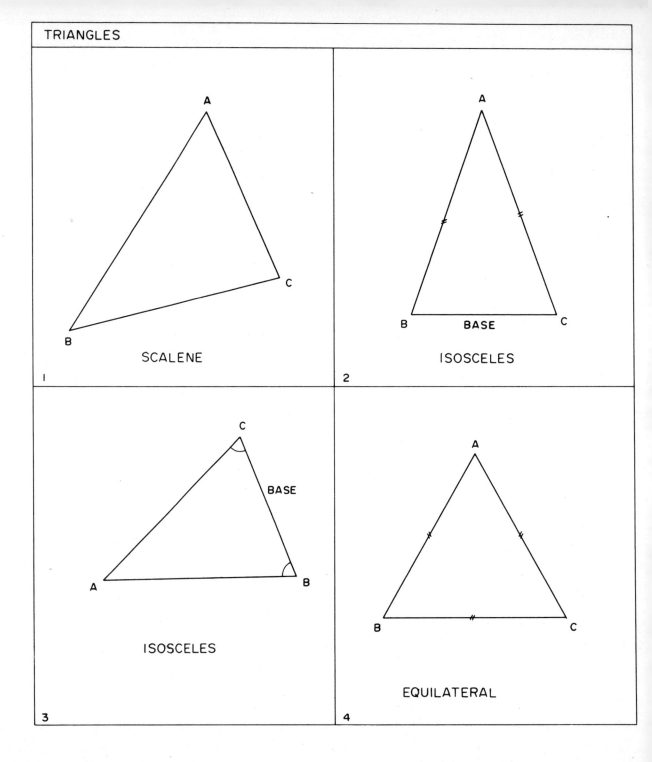

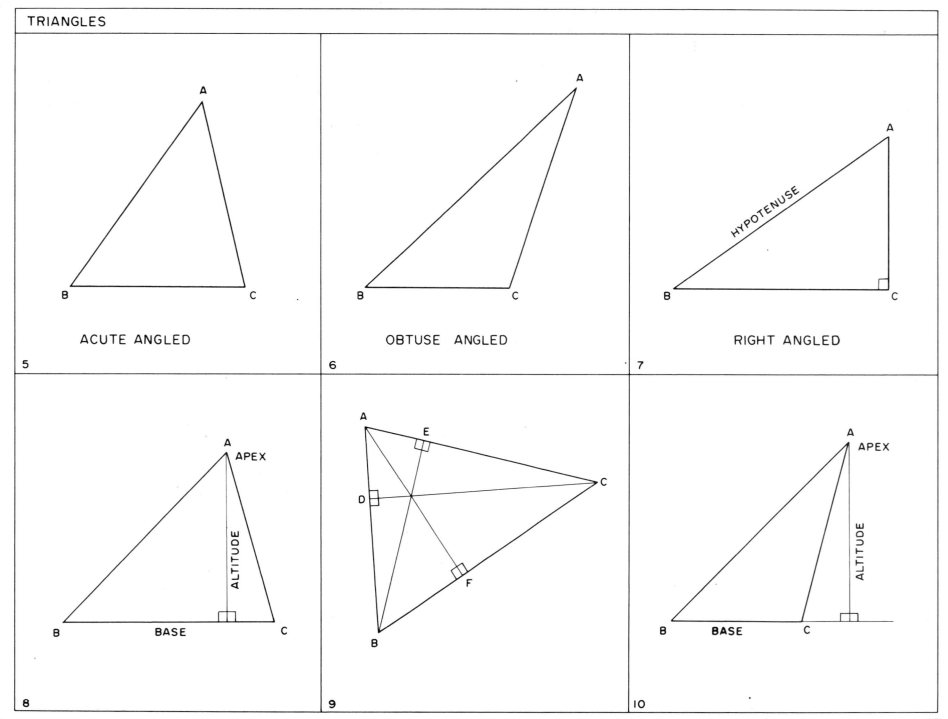

To construct

a) an equilateral triangle with sides of 60 mm

b) an isosceles triangle with a base of 50 mm and two sides of 65 mm each.

c) A right angled triangle with a base of 75 mm and an angle of 30°.

d) a triangle given the length of the three sides.

Panel No:

1. To draw an equilateral triangle with sides of 60 mm first draw the base AB 60 mm long.

2. With your compasses set to 60 mm and their point on A draw an arc EF.

   With your compasses still set to 60 mm and their point on B draw an arc GH to fix point C.

   Draw lines AC and BC to complete the triangle.

3. To draw an isosceles triangle with a base of 50 mm and two sides of 65 mm each.

   Draw the base AB 50 mm long.

4. Set your compasses to 65 mm and with their point on A draw an arc EF.

   Without altering your compasses place their point on B and draw an arc GH to fix point C.

   Draw lines AC and BC to complete the triangle.

5. To draw a right-angled triangle with a base of 75 mm and an angle of 30°. Draw the base AB 75 mm long.

6. At A draw a line AC perpendicular (at 90°) to the base AB.

7. From B draw a line at 30° to the base AB to intersect the line AC.

8. To draw a triangle with sides of 80 mm 60 mm and 100 mm.

   Draw the base AB 80 mm long.

9. Set your compasses to 60 mm and with their point at A draw the arc EF.

10. Now set your compasses to 100 mm and with their point at B draw the arc GH to fix point C.

    Draw AC and BC to complete the triangle.

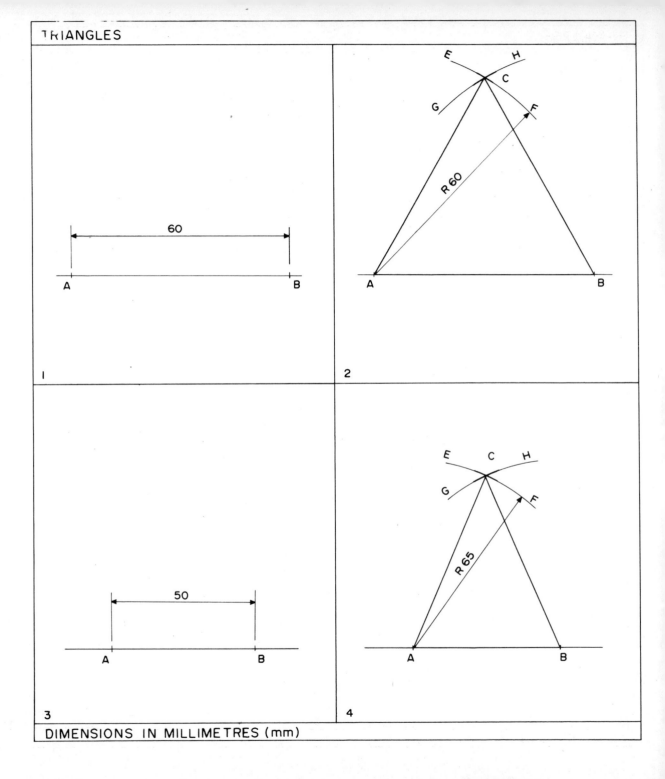

DIMENSIONS IN MILLIMETRES (mm)

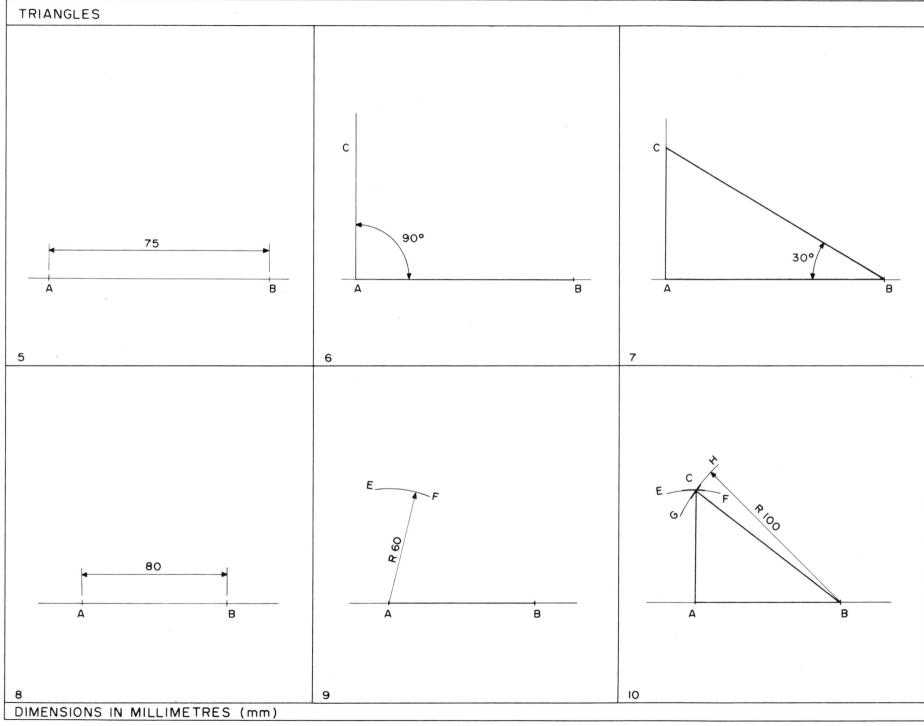

To construct

a) a triangle given (i) the base, (ii) one base angle (iii) the altitude.

b) A triangle given (i) the base (ii) one base angle (iii) the angle at the apex.

c) an isosceles triangle given (i) the base (ii) the altitude.

Panel No:

1. To draw a triangle with a base of 70 mm, a base angle of 60°, and an altitude of 55 mm.

   First draw the base AB 70 mm long.

2. At A draw a line AD at 90° to the base AB and measure along it the altitude AD = 55 mm.

3. Draw line DE parallel to AB.

4. From A draw a line at 60° to intersect line DE at C.

5. Draw sides AC and BC to complete the triangle.

6. To draw a triangle with a base of 70 mm a base angle of 52° and an angle at the apex of 43°.

   First draw the base AB 70 mm long.

7. At A draw a line AC at 52° to the base AB. Use a protractor to measure the angle of 52°.

8. The angle at the apex (C) is 43° but the position of C is not known. However, the angle at B is 85° (180° - (52° + 43°)) so from B draw a line at 85° to the base AB to fix point C. Line in sides AC and BC to complete the triangle.

9. To draw an isosceles triangle with a base of 60 mm and an altitude of 75 mm.

   First draw the base AB 60 mm long.

   Now bisect the base and draw the perpendicular bisector CD.

10. Measure the altitude (75mm) along the line DC to fix C.

    Draw sides AC and BC to complete the triangle.

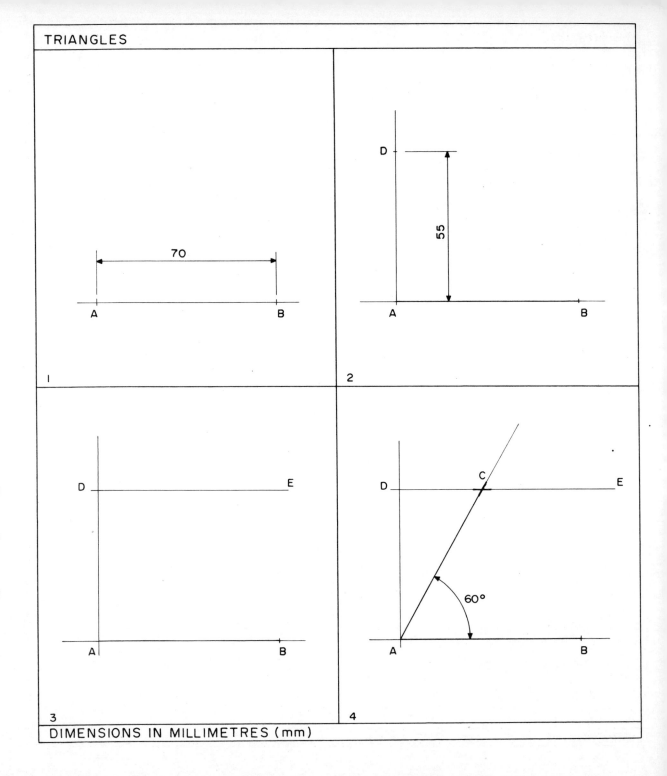

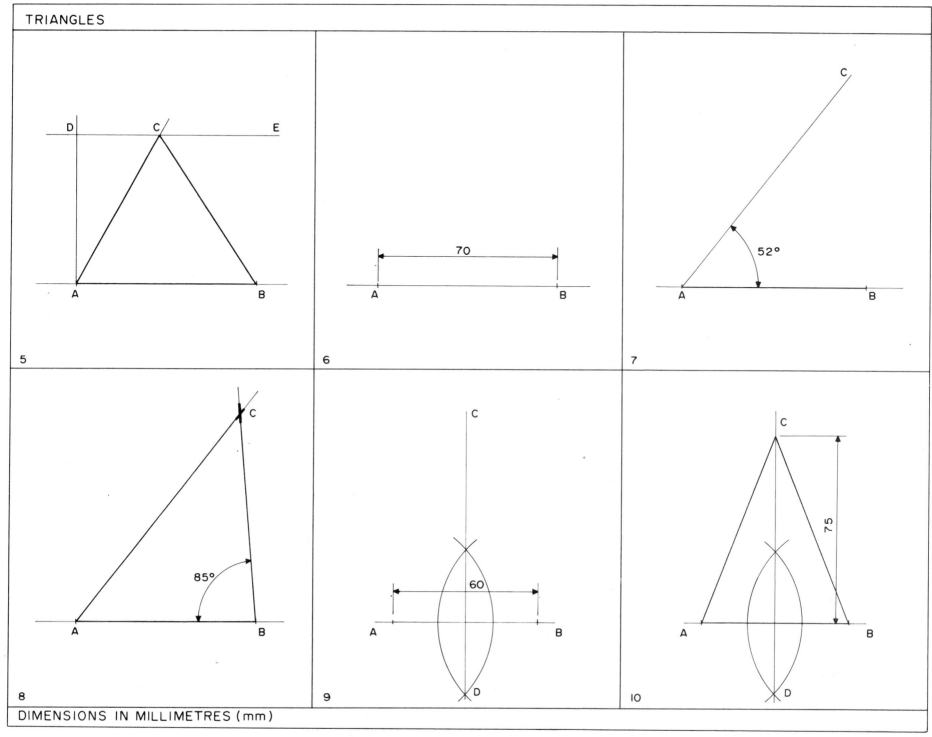

DIVISION OF LINES

To divide a line 100 mm long into ten equal parts would be a simple exercise with a ruler. To divide it into seven equal parts, however, a geometrical method would be more convenient.

To divide a line 100 mm long into seven equal parts.
Draw this exercise following, stage by stage, the instructions and the drawings in the panels.

Panel No:

1 Draw line AB 100 mm long.

2 Draw line AC at any angle to AB.

3 With your compasses mark off seven equal units of any measurement along the line AC.

4 Join point No. 7 to B

5 Using your set square and a straight edge draw lines parallel to 7B through the remaining six points.

6 The construction you have just used relies upon the properties of similar triangles i.e., all angles are equal and the sides are in proportion. In the triangles ABC and DEF which are similar the side DF is three times the length of side AC: The side DE is three times AB and EF is three times BC.

To divide a line 95 mm long into two parts in the ratio of 1 : 2.
Draw this exercise following, stage by stage, the instructions and the drawings in the panels.

Panel No:

7 Draw a line AB 95 mm long and draw line AC at any angle to it.

8 With your compasses mark off three equal units (of any size) along AC.

9 Join point No. 3 to B and parallel to 3B draw line 1D.

AD : DB = 1 : 2

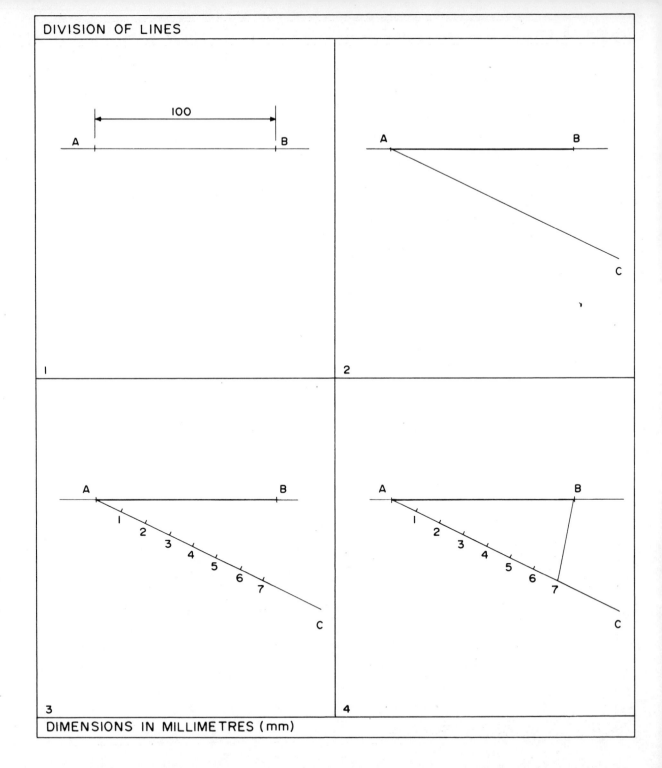

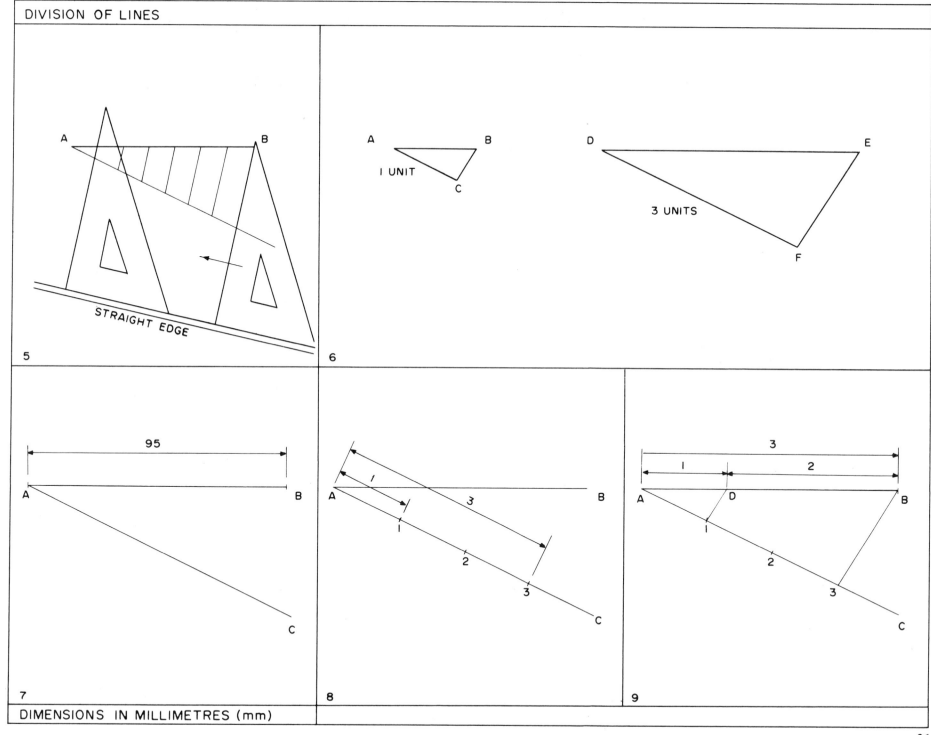

To divide a line 106 mm long into three parts in the ratio of 2 : 3 : 5.

Draw this exercise following, stage by stage, the instructions and the drawings in the panels.

Panel No:

1. Draw a line AB 106 mm long and from A draw a line AC at any angle.

2. As we are dividing in the ratio of 2 : 3 : 5, we add these numbers together 2 + 3 + 5 = 10. Now step off along line AC ten equal divisions.

3. Two units along AC mark point D. A further three units along mark point E and a further five units along mark point F.

4. Join point F to point B with a straight line and parallel to this draw lines through points E and D.

   Line AB is now divided in the ratio of 2 : 3 : 5.

To draw a rectangle, whose length is 105 mm, with its sides in the same ratio as a rectangle 90 mm by 50 mm.

Panel No:

5. Draw a line AB 105 mm long and from A draw a straight line AC at any angle.

6. From A along AC mark off AD which is 90 mm (equal to the length of the given rectangle).

7. From A along AC mark off AE which is 50 mm (equal to the width of the given rectangle).

8. Join D to B with a straight line.

9. Parallel to DB draw a line through point E to give point F in line AB.

10. As AF is the width of the required rectangle it can now be drawn.
    Measure and state the length of FB.

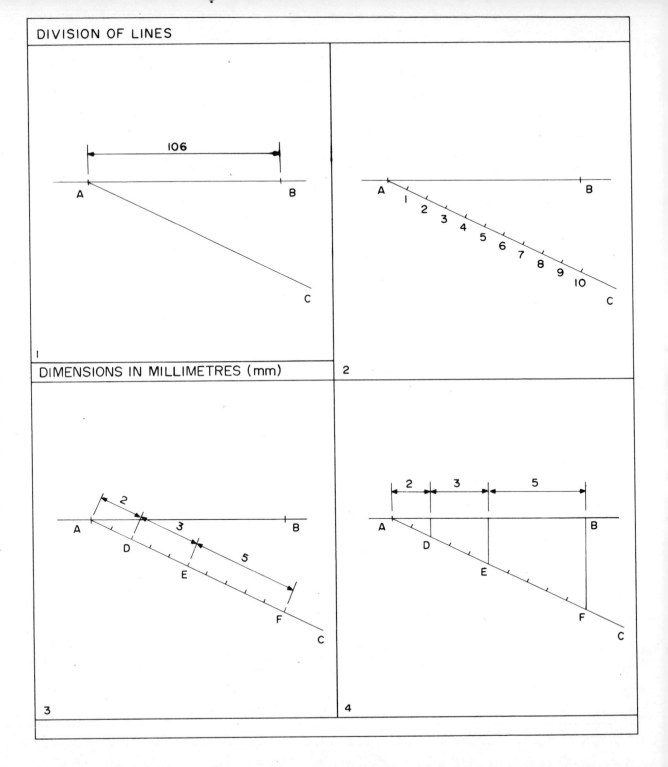

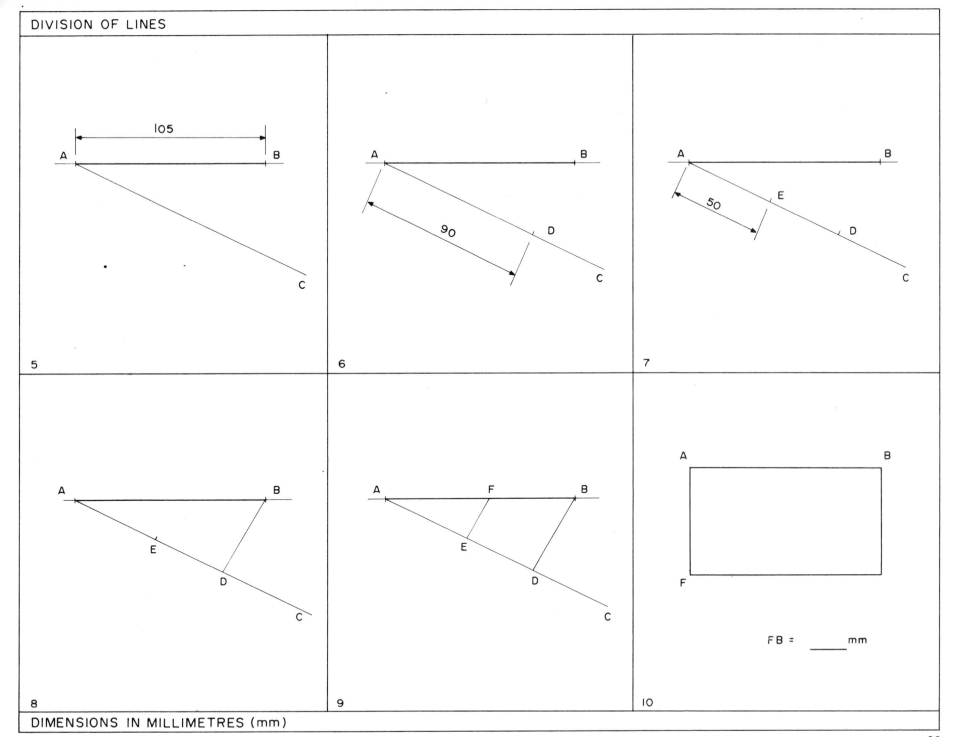

SCALES

If an object is either too large or too small to be drawn full size then the drawing is reduced or enlarged by means of a scale. A scale can be described in words or shown as a fraction called the representative fraction, e.g.:-

    Twice full size or R.F. 2/1
    Half full size or R.F. 1/2

To construct an open divided plain scale of twice full size to measure in millimetres up to a maximum of 100 mm.

Draw this exercise following, stage by stage, the instructions and the drawings in the panels.

Panel No:

1. Draw a line 200 mm long and divide it into portions of 20 mm.

2. Mark the width of the scale 20 mm and also draw a faint line at 10 mm.

3. Line in the scale as shown.

4. Divide the first unit into ten equal parts using your ruler (each part 2 mm).

5. Complete the numbering of the scale as shown. Compasses or dividers are used to take measurements from the scale to a drawing and two measurements are indicated for you.

    Compare the manner in which the scale is numbered to the way in which an ordinary ruler is numbered. Why the difference?

6. Using the scale you have just constructed draw the right angled triangle ABC

        AB = 80 mm.   AC = 50 mm.

    Measure and state the length of CB.

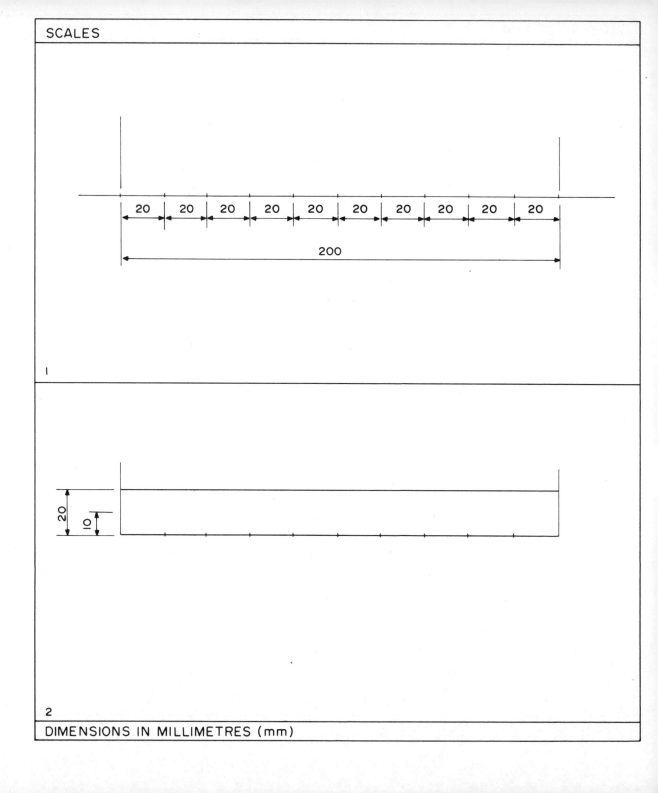

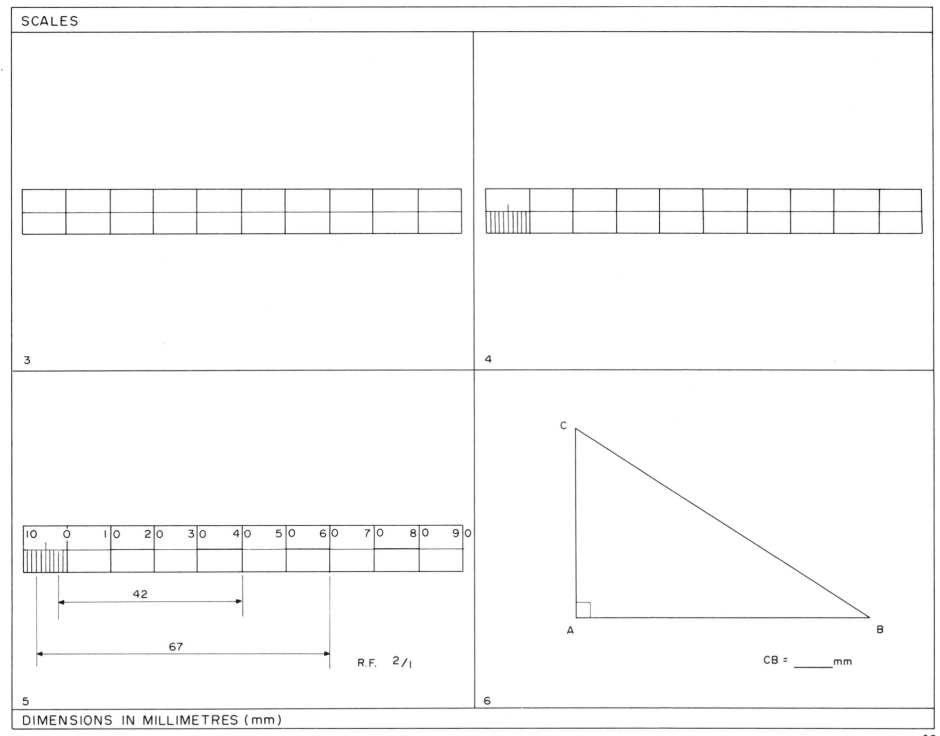

DIAGONAL SCALES

If a plain scale is to be used to read small fractions of a unit it can be difficult to draw and to read. The diagonal scale is an attempt to overcome this.

Panels Nos. 1 to 4 attempt to show the principles used in diagonal scales.

Panel No:

1. In the triangle ABC the line DE bisects the side BC. As you can see DE is, therefore, half of length AB. (Similar triangles)

2. In the triangle ABC the line GF is drawn one tenth the distance along side CB. As you can see GF is, therefore, one tenth of length AB.

3. The two triangles ABC are identical and it can be seen clearly that while it is easy to divide BC into ten parts, it is almost impossible to do the same to side AB.

4. This shows how a small unit can be divided into ten equal parts quite accurately.

To construct a diagonal scale one and a half times full size (R.F. 3/2) to measure in millimetres and to read to one place of decimals.

Draw this exercise following, stage by stage, the instructions and the drawings in the panels.

Panel No:

5. Draw a straight line and mark on it units of 1.5 cm.

6. Complete the rectangle as shown making the height convenient for division into ten with a ruler (40 mm is suggested).

7. Divide the first unit into ten equal parts.

8. Draw faint vertical lines as shown.

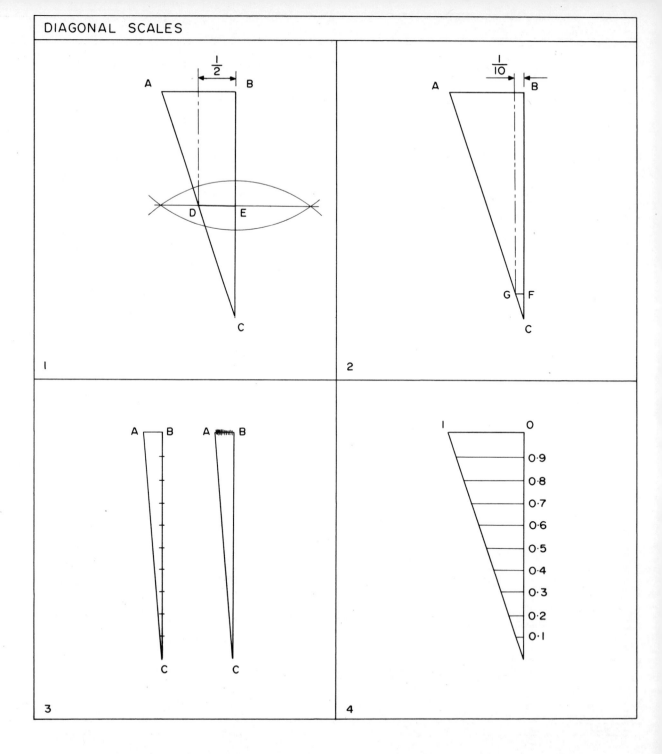

DIAGONAL SCALES

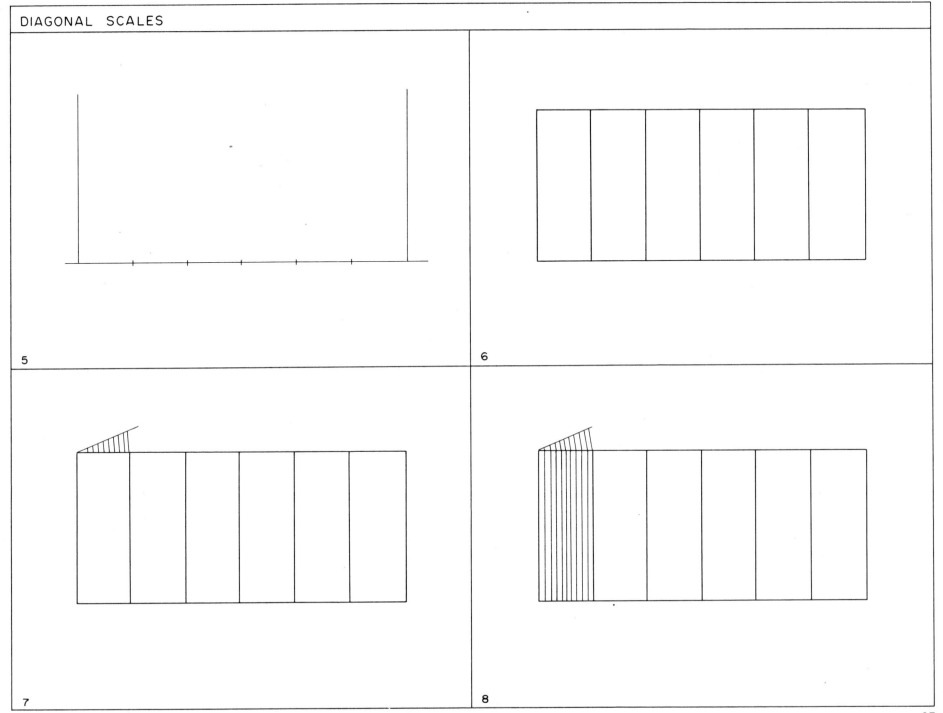

9 The next four panels show the first unit enlarged. Draw in the first diagonal as shown.

10 This shows the next diagonal drawn in.

11, 12 Complete the drawing of the diagonals as shown.

13 Mark ten equal divisions up the left hand side of the scale.

14 Draw horizontal lines dividing the height of the scale into ten.

15 This shows the scale in use to measure millimetres.

16 This shows the scale in use to measure millimetres and tenths of millimetres.

Exercise:
Use your scale to draw a triangle ABC where AB = 45.3 mm, BC = 58.7 mm, and AC = 36.2 mm.

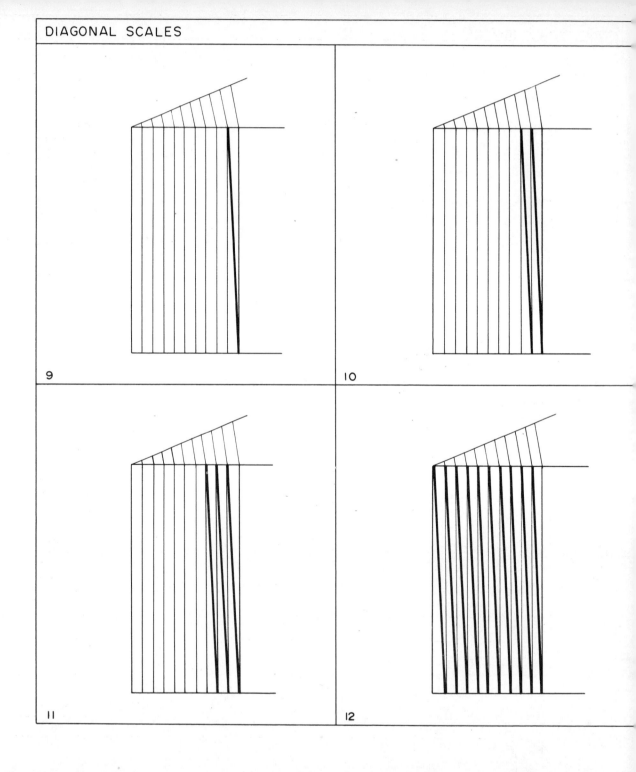

DIAGONAL SCALES

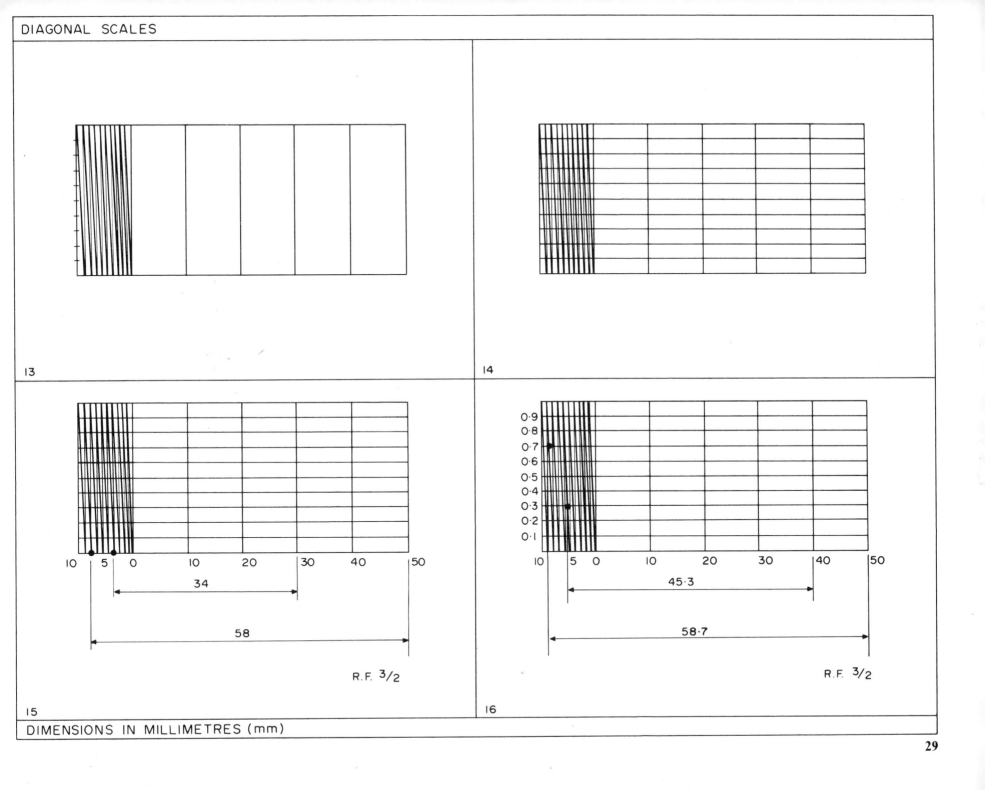

REGULAR POLYGONS

Regular polygons are plane figures whose sides and angles are all equal.

Panels Nos. 1, 2, 3 and 4 show a regular pentagon, hexagon, heptagon and octagon.

To draw a regular hexagon given the length of one side.

Draw this exercise following, stage by stage, the instructions and the drawings in the panels.

Panel No:

5   This shows a regular hexagon with side of 35 mm.

6   With your compasses set to 35 mm (length of side of the hexagon) draw a faint circle.

7   Using your 60°/30° set square and your tee square draw in one side as shown.

8, 9, 10   These panels show how the 60°/30° set square is used to draw three more sides.

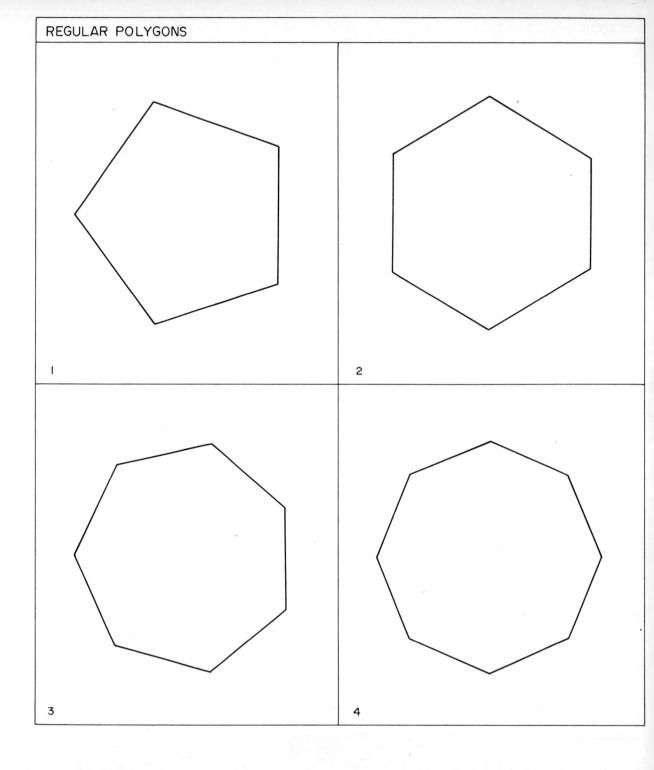

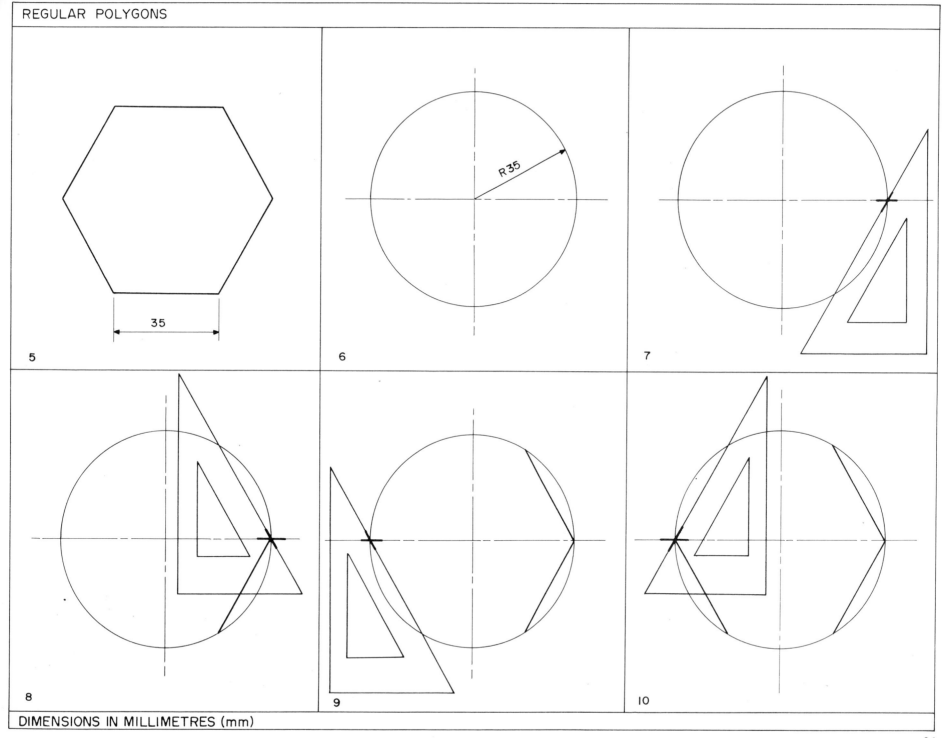

11 You should now have drawn four sides of the hexagon.

12 Complete the hexagon by drawing in the two remaining sides with your tee square.

13 This shows the same hexagon turned through 90°.

14, 15 Draw a 35 mm radius circle as before and use your 60°/30° set square as shown to draw four of the hexagon's sides.

16 You should now have four sides of the hexagon drawn.

17 Using a set square and tee square draw the two remaining vertical sides of the hexagon.

To draw a regular hexagon with ruler and compasses only.

Panel No:

18, 19 Draw a faint circle whose radius is equal to the length of one side of the required hexagon.

Starting at A you can now step off the radius of the circle six times around the circumference.

20 Join the six points with straight lines to form the hexagon.

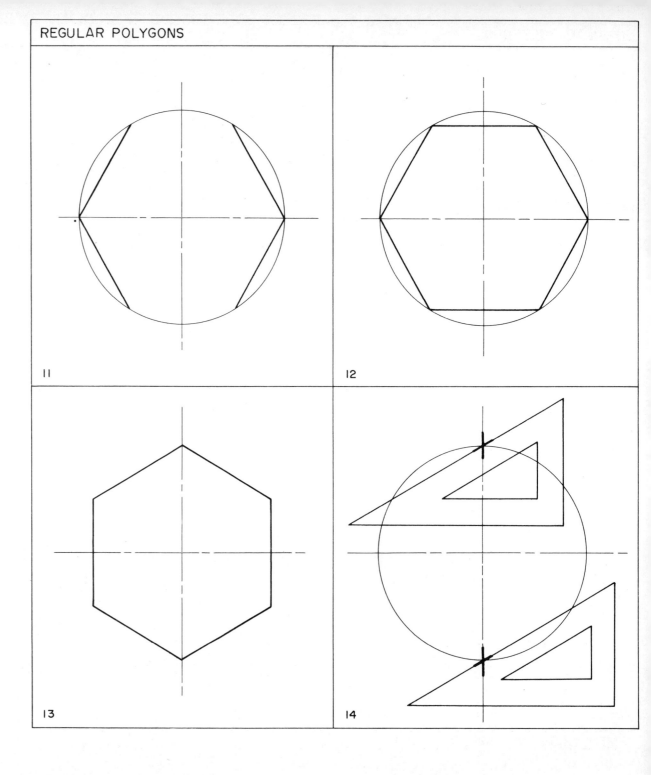

REGULAR POLYGONS

REGULAR POLYGONS

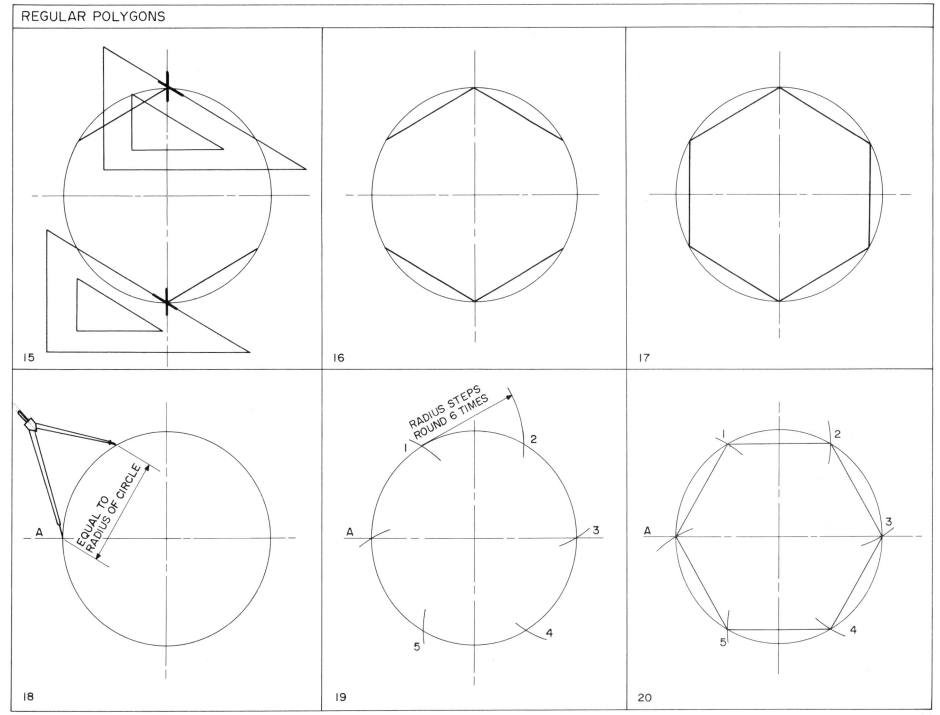

In Engineering Drawing it is often necessary to draw hexagons when the distance across the FLATS (A/F) is given.

To draw a regular hexagon which is 50 mm A/F.
Draw this exercise following, stage by stage, the instructions and the drawings in the panels.

Panel No:

1 This shows the required hexagon.

2 Draw a circle of 50 mm diameter (equal to the distance A/F).

3 Using your tee square and 60°/30° set square as shown, draw two faint lines to touch the circle.

4 You should now have two faint lines touching the circle as shown.

5 Reversing the set square draw two more faint lines, touching the circle, as shown.

6 You should now have four faint lines as shown.

7 Using your tee square and set square as shown draw two faint vertical lines touching the circle.

8 You should now have six faint lines touching the circle as shown.

9 This panel shows the completed hexagon.

Exercise:

10 Draw the two given hexagons to the dimensions shown.

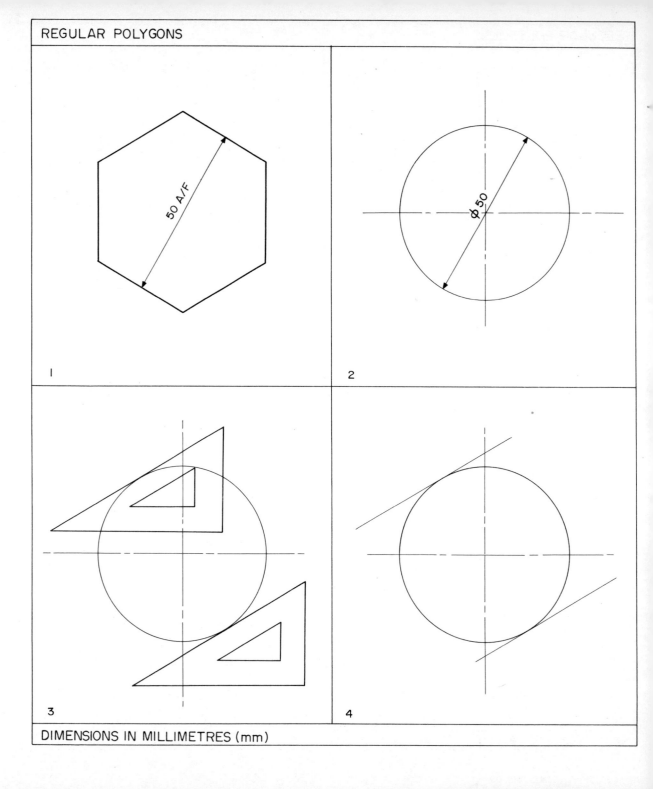

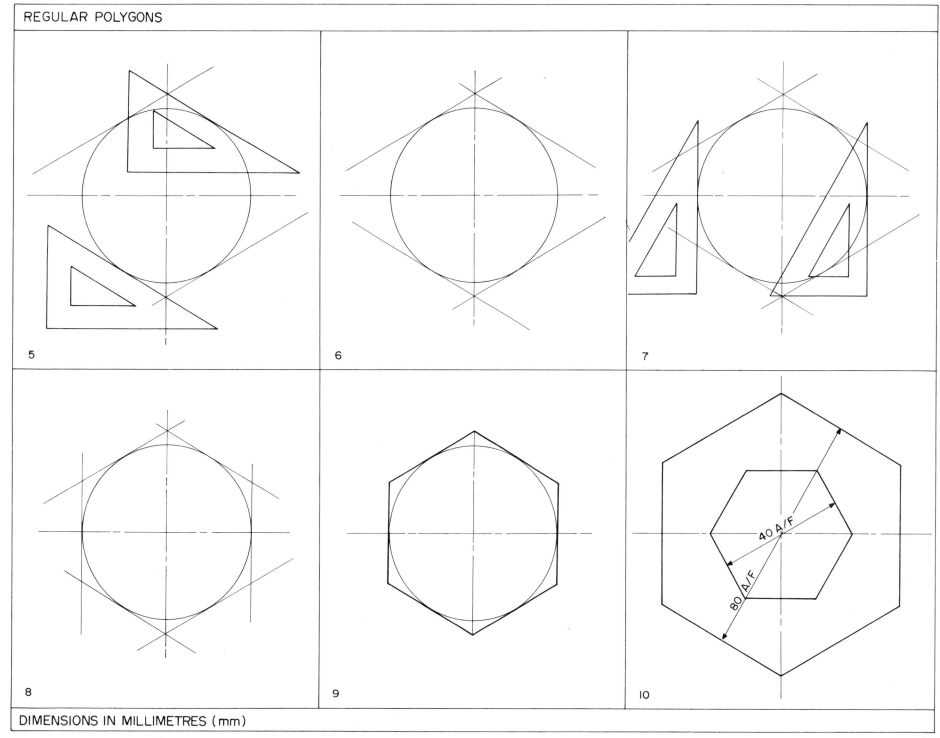

The octagon has fewer applications than the hexagon but its construction is worth learning.

To draw a regular octagon which is 50 mm A/F (across flats).

Draw this exercise following, stage by stage, the instructions and the drawings in the panels.

Panel No:

1. Draw a circle of 50 mm diameter.

2. Using your tee square and 45° set square as shown draw two faint lines touching the circle.

3. Slide the set square to draw the two further lines touching the circle.

4. Using your tee square and set square as shown draw two further lines to touch the circle and at right angles to the tee square.

5. Complete the octagon drawing the last two lines with your tee square.

6. Line in the required octagon.

To draw a regular octagon within a given circle.

Panel No:

7. Draw a circle of 60 mm diameter with centre lines as shown.

8. Using your 45° set square draw two more centre lines as shown.

9. Join the intersections on the circumference as shown.

10. This panel shows the completed octagon.

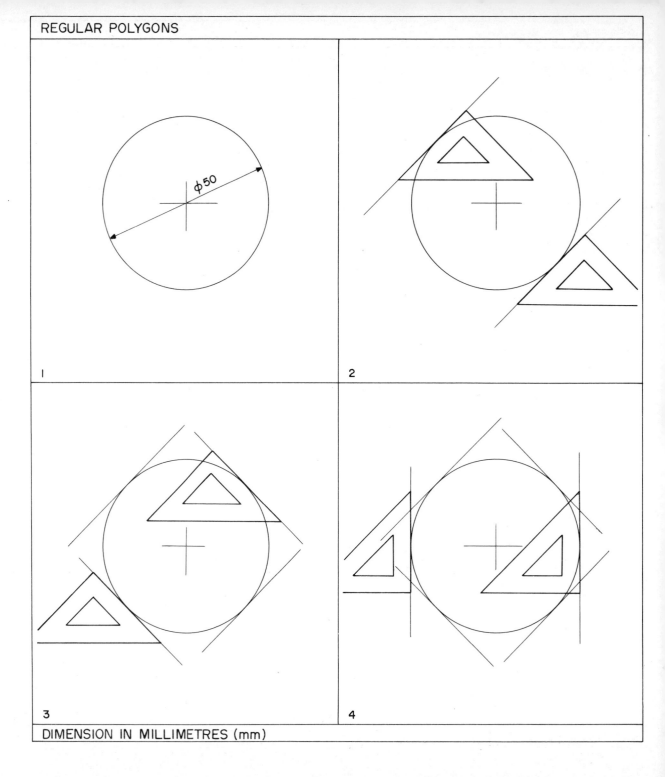

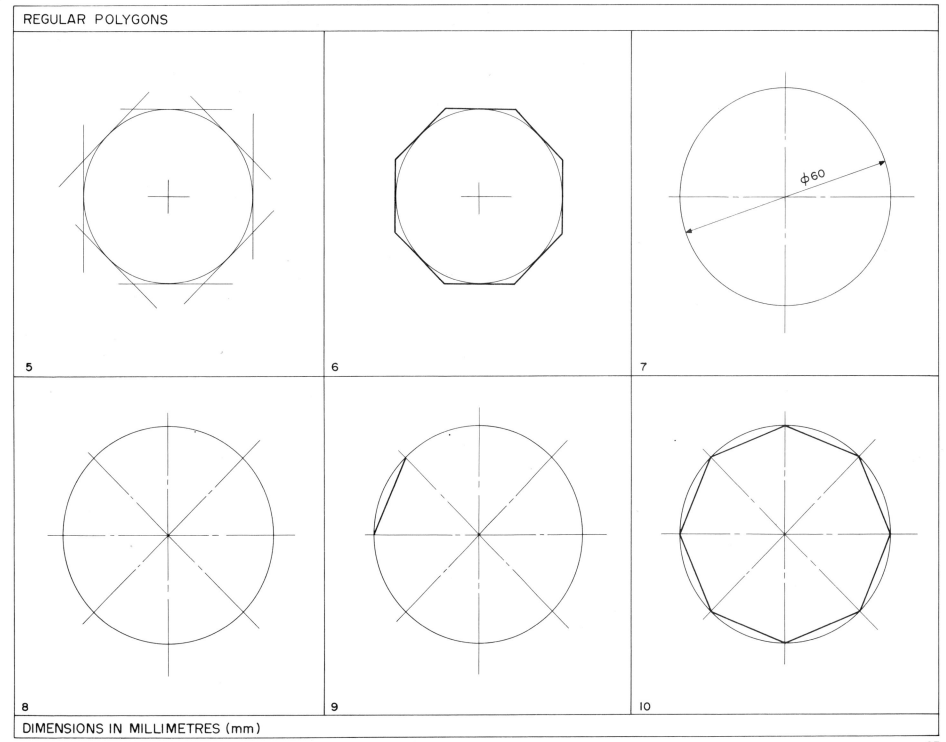

To draw a regular heptagon (seven sides) within a given circle.

Draw this exercise following, stage by stage, the instructions and the drawings in the panels.

Panel No:

1. This shows the required heptagon within a circle of 130 mm diameter.

2. With centre O draw a 130 mm diameter circle. Mark the horizontal diameter AB.

3. Divide the horizontal diameter AB into seven equal parts.
   (Note: If you were drawing a nine sided figure you would divide AB into nine equal parts, a ten sided figure ten parts, etc.).

4. With your compasses set to AB (130 mm) and with A as centre, draw an arc.

   With your compasses still set to AB and with B as centre, draw another arc.

   The intersection of these arcs fixes point C.

5. This shows how your drawing should now look.

6. Through point C and point No. 2 on diameter AB draw a straight line.

   The intersection of this straight line with the circumference fixes point D.
   Note: No matter which polygon you are drawing the line always goes from C through point No. 2 on line AB.)

7. Join A to D with a straight line. This is one side of the required heptagon.

8. Set your compasses to length AD. This length should now step round the circumference seven times.

9. Join the seven points with straight lines to complete the drawing.

10. This shows how your completed drawing should look.

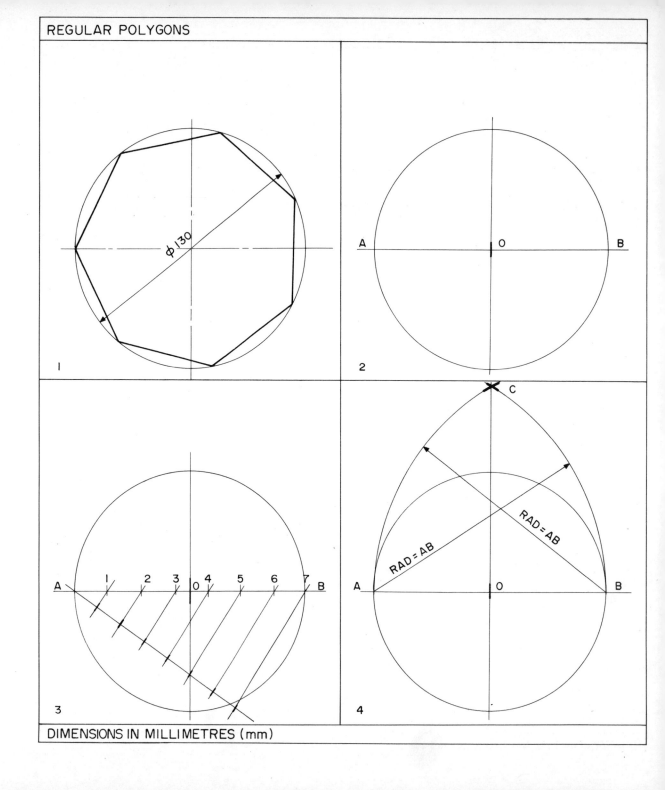

REGULAR POLYGONS

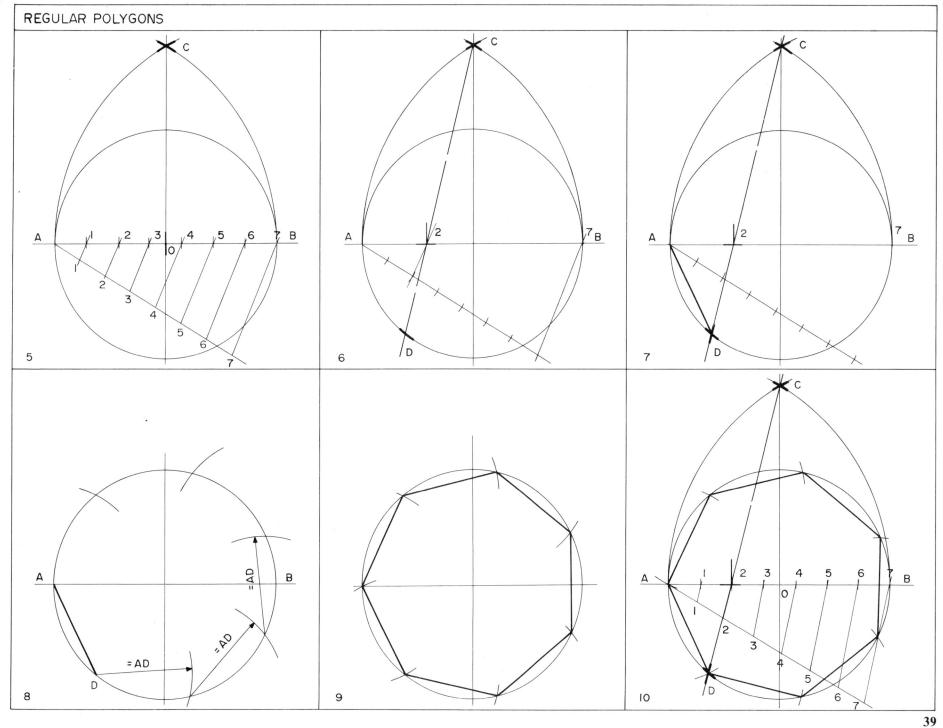

To draw a regular pentagon (five sides) on a given line.

Draw this exercise following, stage by stage, the instructions and the drawings in the panels.

Panel No:

1. Draw the given line AB.

2. Draw the perpendicular bisector of AB.

3. From A draw a line at 45° to cut the bisector at point 4.

4. From B draw a line at 60° to cut the bisector at point 6.

5. Find the mid point between 4 and 6 by drawing the perpendicular bisector to give point 5.

6. With centre 5 draw a circle passing through A and B.

7. Set your compasses to length AB. This length should step round the circumference five times.

8. Join the five points with straight lines to complete the drawing.

9. This shows the completed drawing.

10. By marking the distance 4 to 5 from 6 to give point 7 the construction can be used to draw a heptagon.
    (Note: The construction is an approximate one but does give near perfect results with careful drawing.)

Exercise:

Construct a regular heptagon (seven sides) on a line of 45 mm.

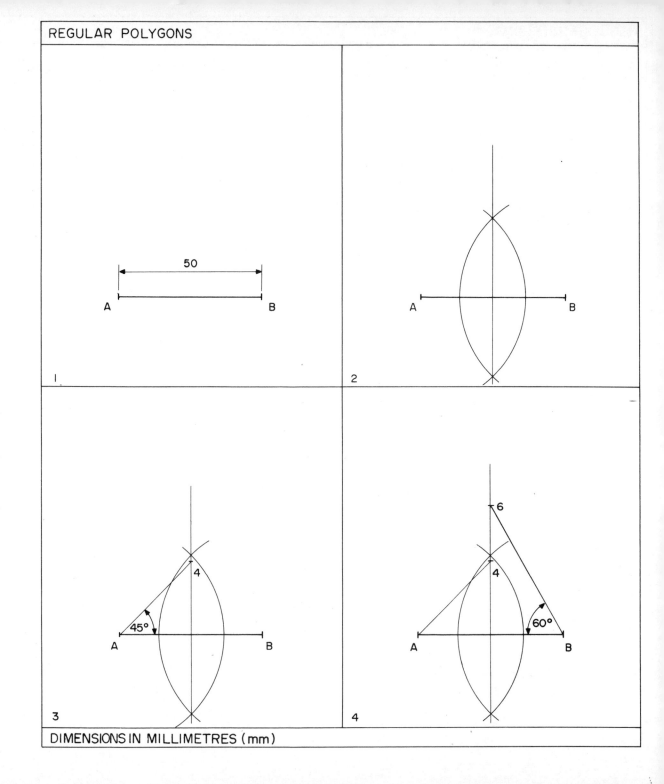

REGULAR POLYGONS

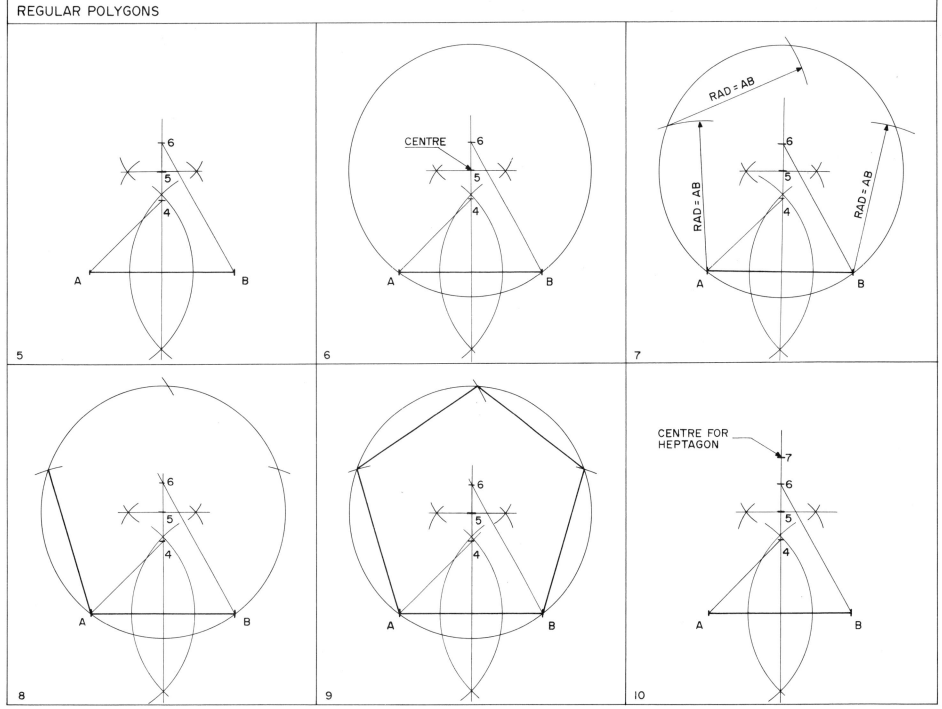

EQUIVALENT AREA

To draw a rectangle equal in area to a given triangle.
Draw this exercise following, stage by stage, the instructions and the drawings in the panels.

Panel No:
1. Draw the given triangle to the given dimensions.
2. Draw CD the height or altitude of the triangle.
3. Bisect CD.
4. Draw AF and BE at 90° to the base AB to complete the rectangle.

   Rectangle ABEF is equal in area to triangle ABC.
   (Note: The above constructions uses the fact that the area of a triangle is $\frac{\text{Base x height}}{2}$.)

To draw an isosceles triangle equal in area to a given triangle and using the same base.

Panel No:
5. This illustrates the fact that triangles on the same base and having the same height are equal in area.

   The area of both these triangles is:
   $$\frac{80 \text{ mm} \times 90 \text{ mm}}{2} = 3\,600 \text{ mm}^2$$

6. Draw this triangle to the given dimensions.
7. Draw ED through the apex of the triangle and parallel to the base AB.
8. Bisect the base AB.

   The intersection of the bisector and ED fixes point F.

9. Draw sides AF and BF.

   Isosceles triangle ABF is equal in area to scalene triangle ABC.

Exercise:
10. Draw a rectangle equal in area to the given equilateral triangle.

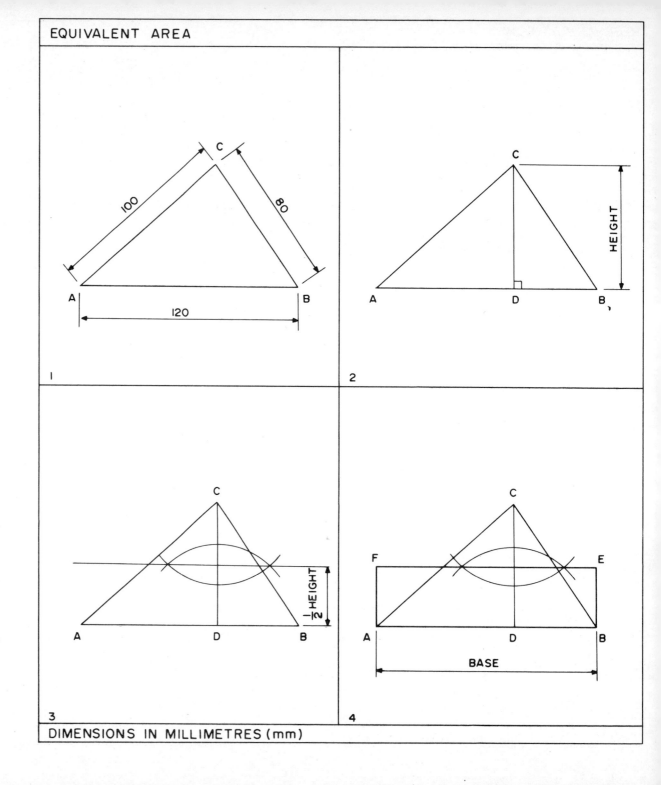

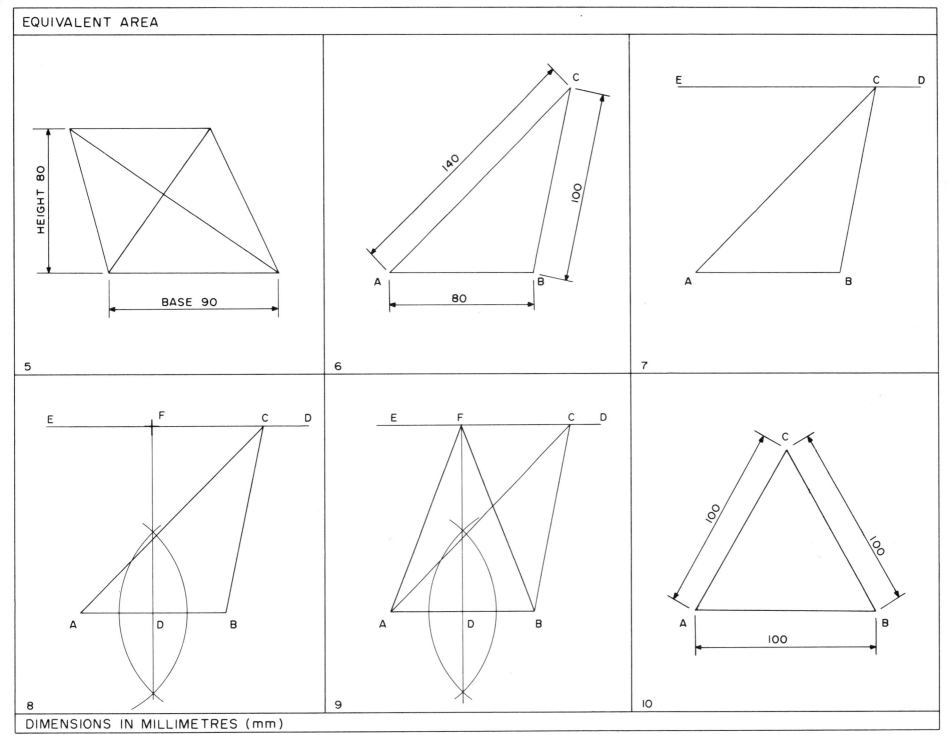

To change a given polygon into a triangle having the same area.

Draw this exercise following, stage by stage, the instructions and the drawings in the panels.

Panel No:

1. Draw the quadrilateral ABCD to the given dimensions.
2. Join DB with a straight line.
3. Parallel to line DB draw a line through C to meet the extension of the base AD at E.
4. Join D to E with a straight line.

   Triangle ADE is equal in area to the quadrilateral ABCD.

To draw a triangle equal in area to a given pentagon.

Panel No:

5. Draw the pentagon ABCDE to the given dimensions.
6. Join D to A with a straight line.
7. Parallel to DA draw a line through E to meet the extension of the base at F.
8. Join D to F.

   Join D to B with a straight line.

9. Parallel to DB draw a line through C to meet the extension of the base at G.
10. Join D to G.

    Triangle FDG is equal in area to the pentagon ABCDE.

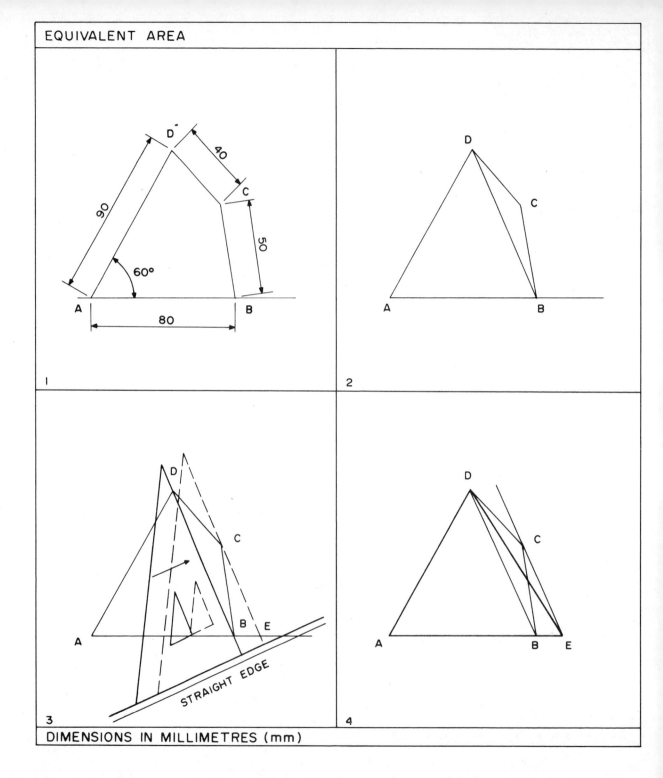

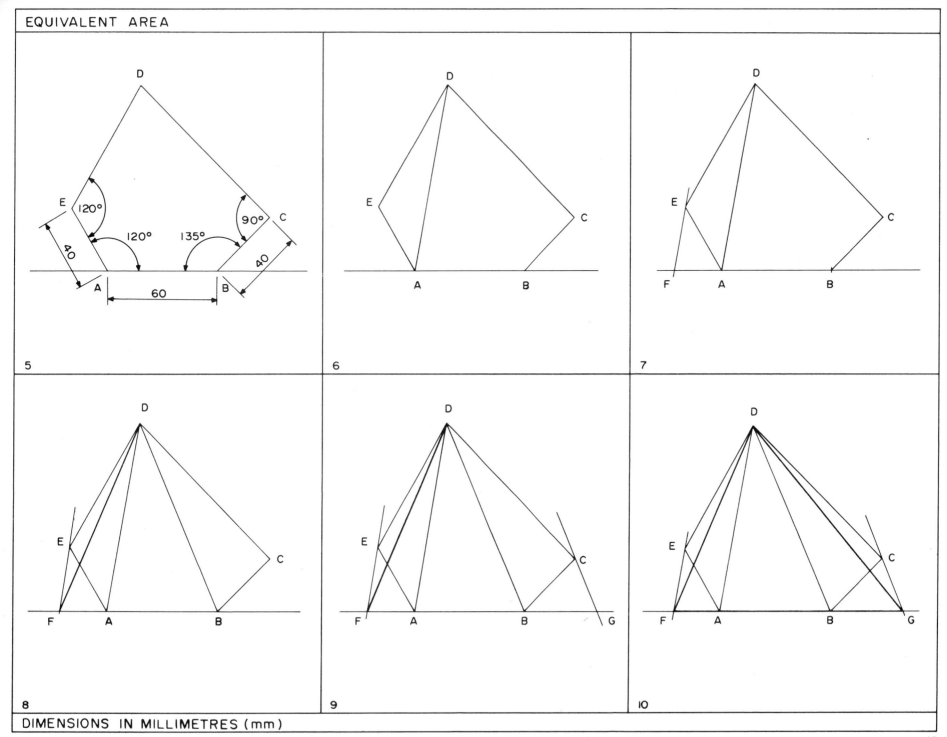

To draw a square equal in area to a given rectangle.
In Panel No. 1 two chords AB and CD intersect at P. It can be proved geometrically that
$$AP.PB = CP.PD$$

Panel No. 2 shows a diameter AB intersecting CD at right angles. CP now equals PD and it follows that $AP.PB = CP^2$.

This is the theory on which the following construction is based. Draw this exercise following, stage by stage, the instructions and the drawings in the panels.

Panel No:

3 Draw the rectangle ABCD to the given dimensions.

4 Produce AB.

5 With the centre B and radius BC locate point E.

6 Draw the perpendicular bisector of AE to fix point F.

7 With F as centre draw the semi-circle, with AE its diameter.

8 Produce CB to intersect the semi-circle at G.
   Note:
   $$AB.BE = BG^2$$

9 Draw the square with length of side equal to BG.

Exercise:

10 Draw a square equal in area to the given rectangle. Use your drawing to measure the $\sqrt{6\,000}$.

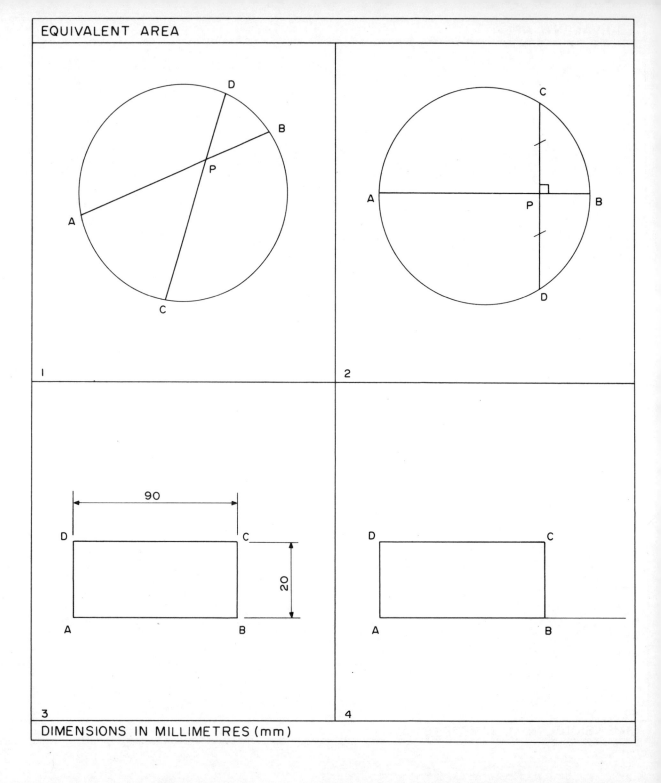

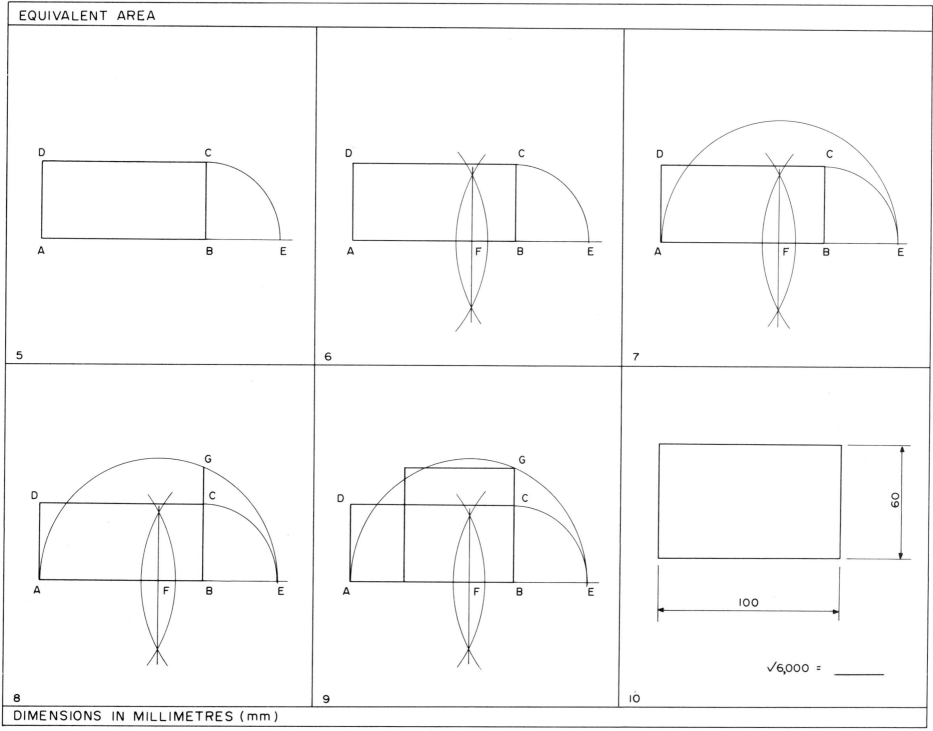

SIMILAR FIGURES

Plane figures are said to be similar if their angles are the same and their sides are in the same proportion.

Panel No:

1. The triangles ABC and DEF are similar. Their angles are the same and EF is twice BC — that is it is in the ratio of 2:1. The other sides are in the same ratio — DF is twice AC and DE is twice AB.

2. Any shape can be drawn similar to another using the method shown in this panel. If the sides of the enlarged figure are to be in the ratio of 3 : 1 then point O must be positioned so that OA:OB is in the ratio of 1:3.

To draw an irregular polygon similar to a given one but with its sides increased in the ratio of 2:1.

Draw this exercise following, stage by stage, the instructions and the drawings in the panels.

Panel No:

3. Draw the irregular pentagon ABCDE to the given dimensions.

4. When drawing similar figures it is usual to set out the enlargement or reduction from one corner of the original figure.

   From D draw radiating lines through each corner.

5. Mark point F so that DF is twice DC.

6. Draw side FG parallel to side CB.

7. Draw side GH parallel to side BA.

8. Complete the enlargement by drawing side HJ parallel to side AE.

   Pentagon FGHJD is similar to the original pentagon but with its sides increased in the ratio of 2:1.

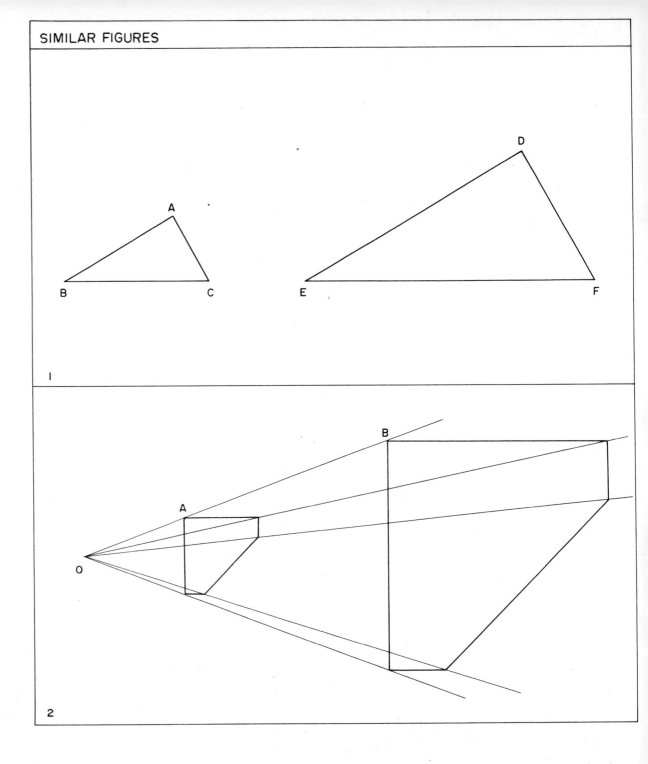

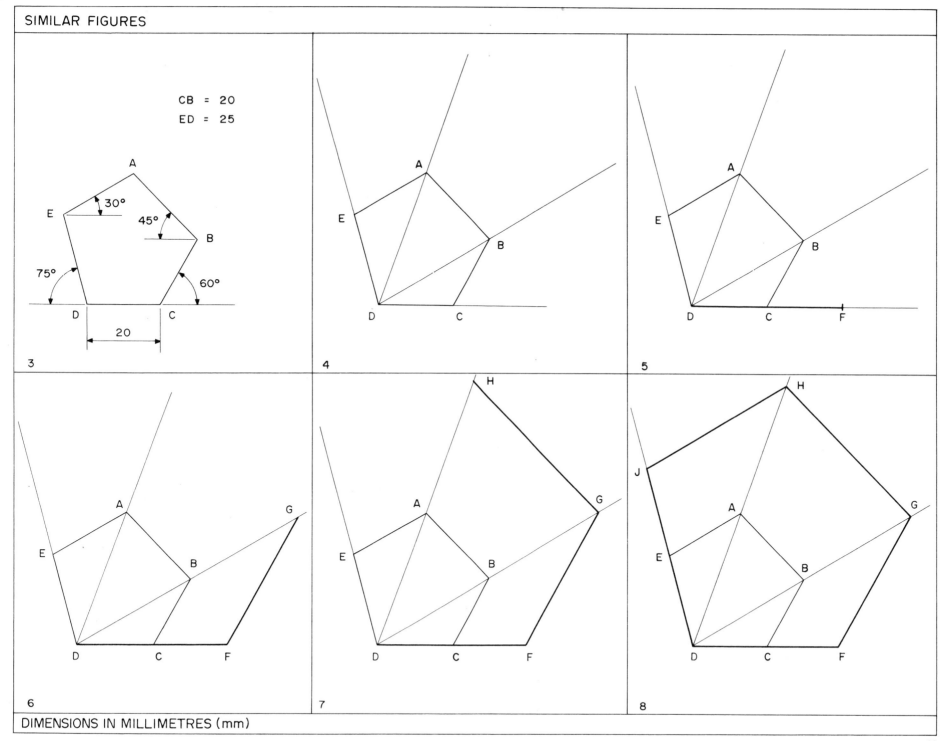

When similar figures are to be drawn with their sides increased or decreased by a ratio that is difficult to set out, the following construction can be used.

To increase a given line in the ratio of 5:3.
Draw this exercise following, stage by stage, the instructions and the drawings in the panels.

Panel No:
1. Draw line AB 56 mm long. From A draw a line AC at any angle.
2. Set out five equal units along AC.
3. Join points 3 and B with a straight line.
4. Parallel to line 3B draw a line through point 5 to fix point D.

   AD:AB is in the ratio of 5:3.

To draw a triangle similar to a given one but with its sides increased in the ratio of 7:5.

Panel No:
5. Draw the triangle ABC to the given dimensions.
6. From A draw line AD to any angle and produce AB.
7. Set out seven equal units along AD.
8. Join points 5 and B with a straight line.
9. Parallel to line 5B draw a line through point 7 to fix point E.
10. Extend side AC and then draw side EF parallel to side BC.
    (Note: AE:AB = EF:BC = AF:AC = 7:5).

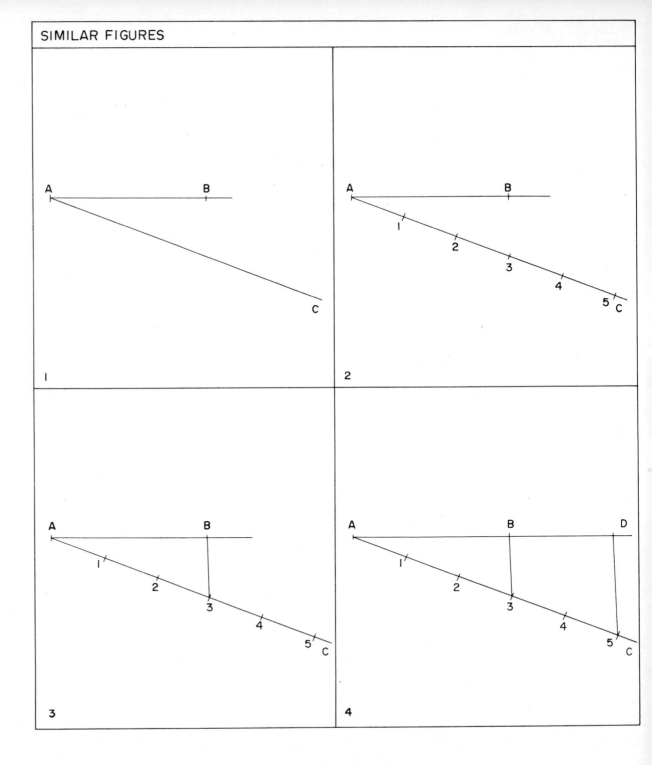

SIMILAR FIGURES

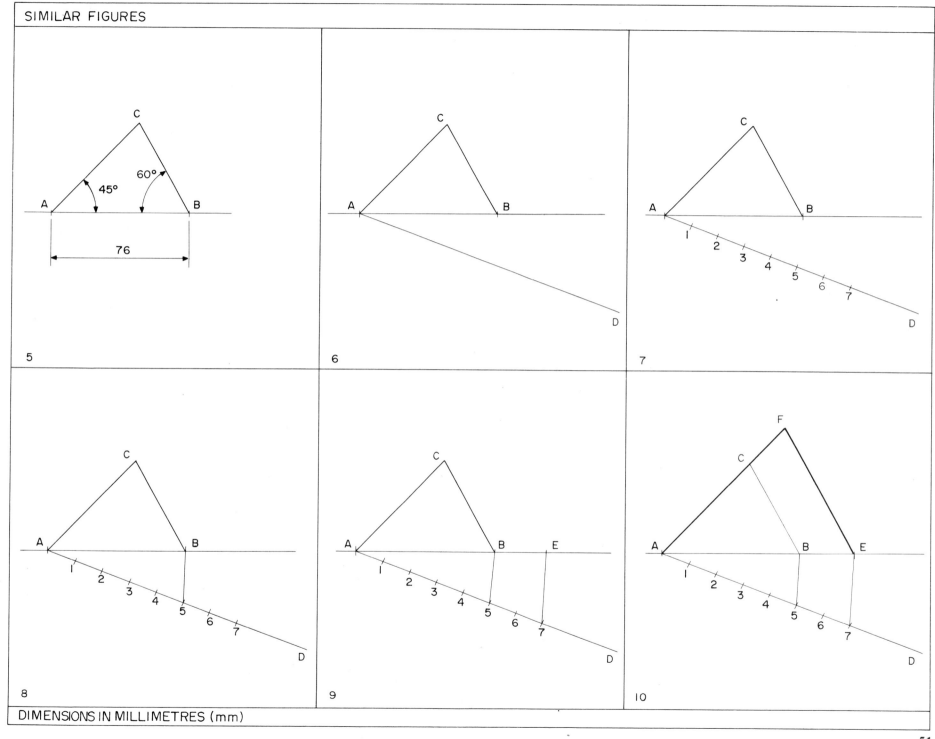

DIMENSIONS IN MILLIMETRES (mm)

To draw a polygon similar to a given one but with its sides reduced in the ratio of 5:6.

Draw this exercise following, stage by stage, the instructions and the drawings in the panels.

Panel No:
1. Draw the pentagon ABCDE to the given dimensions.
2. From corner A, draw radiating lines to the other corners.
3. From A draw line AF at any angle.
4. Set out six equal units along line AF.
5. Join points 6 and B with a straight line.
6. Parallel to line 6B draw a line through point 5 to fix point G.
7. Draw side GH parallel to side BC.
8. Draw side HJ parallel to side CD.
9. Draw side JK parallel to side DE. Pentagon AGHJK is similar to pentagon ABCDE but with its sides reduced in the ratio of 5:6.
   (Note: GH:BC = HJ:CD = JK:DE = KA:EA = 5:6).

Exercise:
10. Draw the given pentagon and then reduce it to a similar figure with its sides in the ratio of 5:8 of the original.

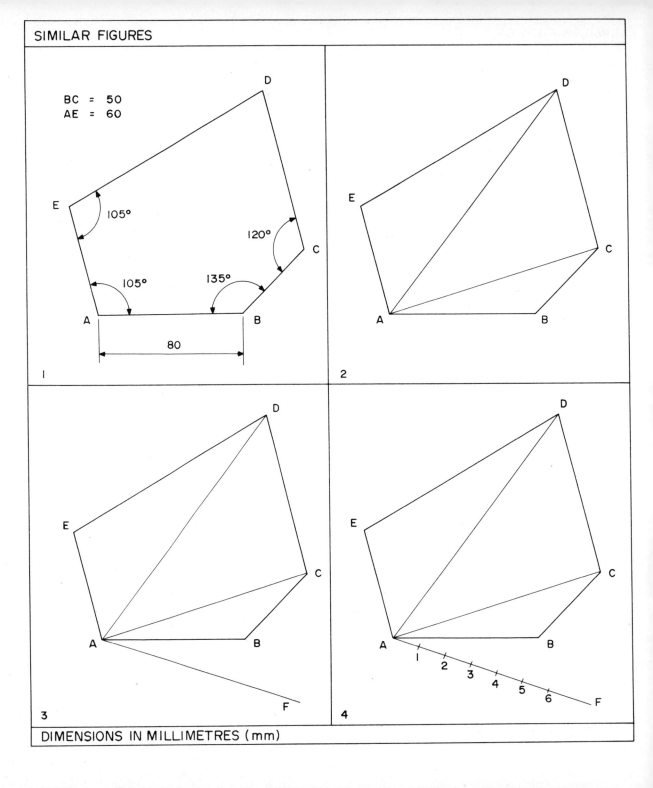

52

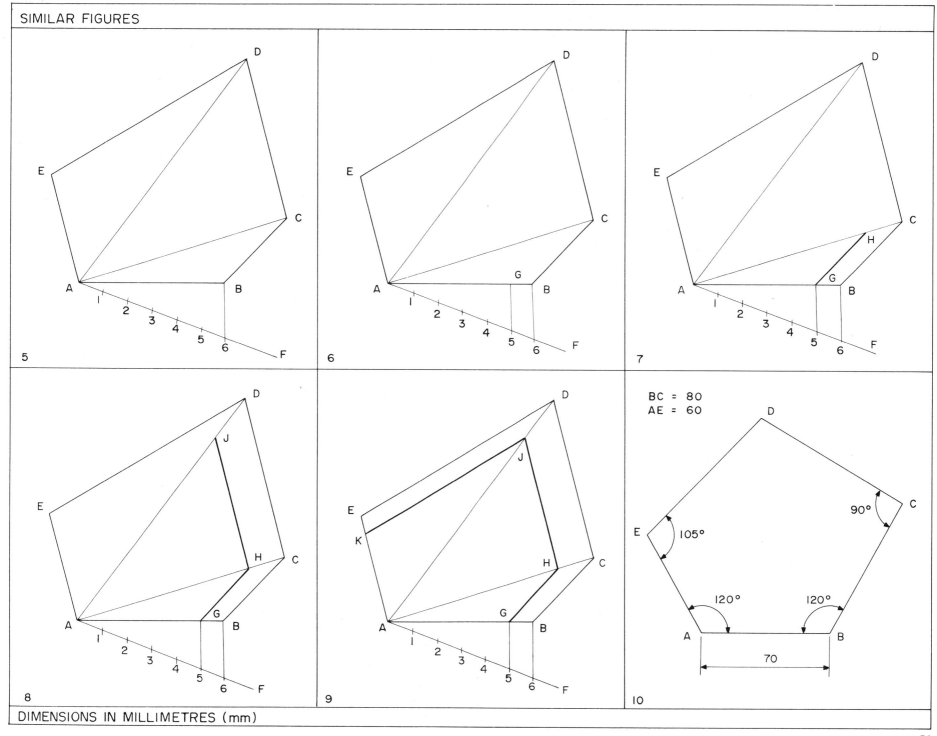

THE CIRCLE

You should make yourself familiar with the names used for various parts of the circle.

Panel No:

1. Draw a circle of 100 mm diameter and neatly letter the parts as shown. The circumference is the complete boundary or perimeter of the circle.

2. Draw another circle of the same size and show on it a chord and a segment.

    A segment is the area enclosed by a chord and an arc.

3. With a radius of 80 mm and centre O draw part of a circle bounded by two radii.

    The area enclosed by the arc (part of the circumference) and the two radii is called a sector.

To draw a circle within a rectangle as shown in Panel 7.

Panel No:

4. Draw the rectangle to the given dimensions.

5. In the position shown draw the centre line (a long dash followed by a short dash — a chain line).

6. Draw the second centre line in the position shown.

    These two centre lines fix the position of the centre of the circle.

7. With your compasses set to 30 mm, draw the required circle.

When a number of circles have their centres on the circumference of another circle this is called a pitch circle and it is drawn as a chain line.

To draw the figure shown in Panel No. 10.

Panel No:

8. Draw centre lines at right angles to fix the position of centre O.

    With your compasses set at 70 mm draw the outer circle.

    With your compasses set at 50 mm draw the pitch circle.

9. Use your 45° set square to fix the centres of the remaining four holes.

10. Draw the eight 20 mm diameter holes.

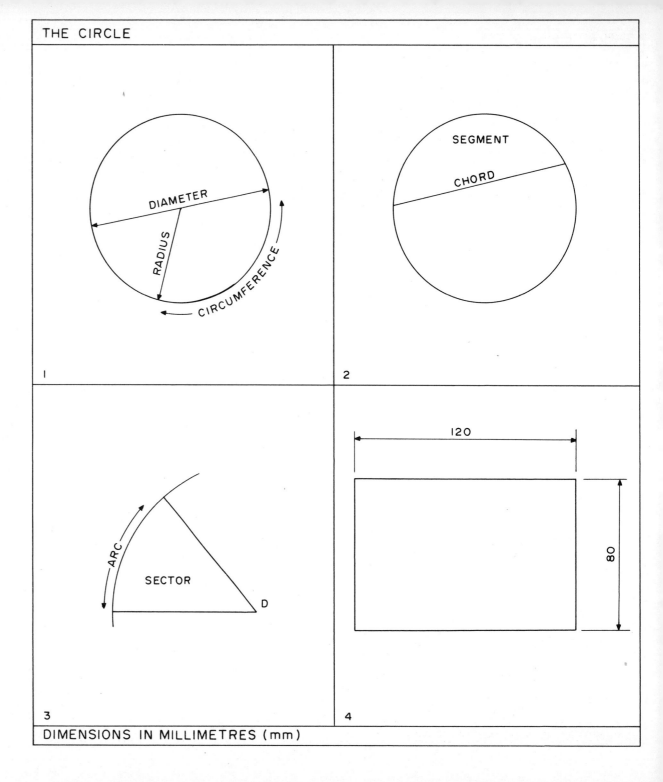

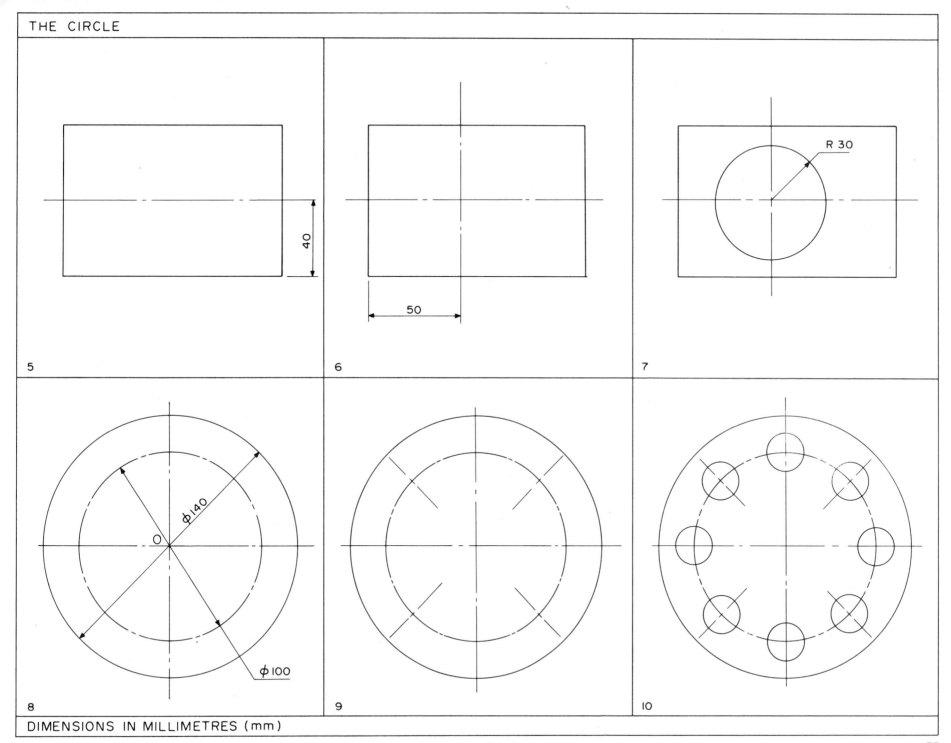

TANGENTS

A tangent is a straight line which touches the circumference of a circle (but does not cut it), at one point only, and makes an angle of 90° with the radius drawn through that point. In Panel No. 1 TN is a tangent to the circle.

To draw a tangent to a circle at a given point on the circumference.
Draw this exercise following, stage by stage, the instructions and the drawings in the panels.

Panel No:

2 Fix the centre of the circle O.

3 With O as the centre draw a circle of 50 mm diameter.

4 Mark a point P on the circumference of the circle.

5 Draw a straight line to pass through P.

6 From P mark PA equal to OP.

7 With any radius greater than OP draw an arc ef with centre A.

8 From centre O and with the same radius draw an arc intersecting ef at D.

9 A straight line drawn through D and P is the required tangent.

10 Draw the given view of the bracket where P marks the points of tangency. Using the construction you have just learned find the points where the straight sides meet the base AB. Measure and write down the length of AB.

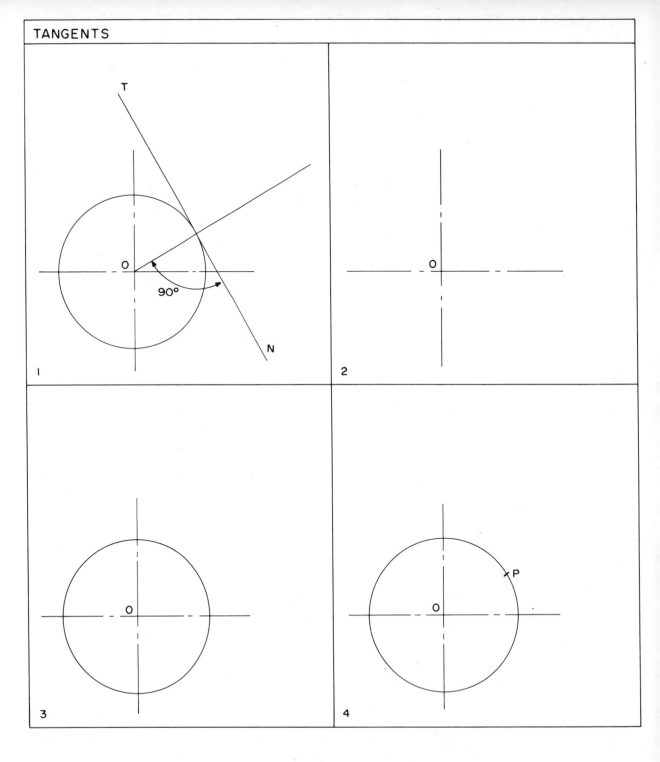

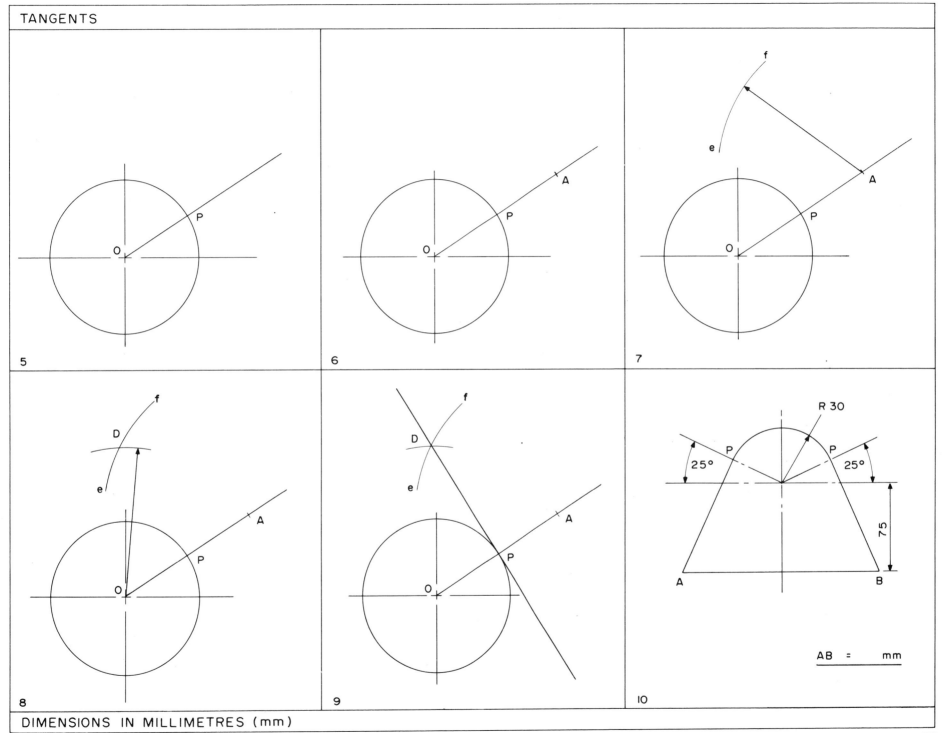

In drawing a tangent to pass through a given point outside the circumference of a circle we make use of the fact that the angle in a semi-circle is 90°. This is illustrated in Panel No. 1. If we take any point such as A on the circumference of a semi-circle and from it draw straight lines — one to each end of the diameter C and D — the resulting angle is a right angle. This applies to any point in the circumference — hence the angle at B is also a right angle.

To draw a tangent to a circle to pass through a point outside the circumference.
Draw this exercise following, stage by stage, the instructions and the drawings in the panels.

Panel No:
2 In this panel details are given of a circle and a point A through which a tangent is to be drawn as follows.

3 Fix the centre of the circle O.

4 Draw the circle and fix point A.

5 Join point A and the centre of the circle O with a straight line.

6 Bisect line OA to find its centre B.

7 With B as the centre and with compasses set to radius BA draw the semi-circle OA. This give you point P where the semi-circle cuts the circumference of the circle.

8 Draw a straight line joining A and P. This is a tangent to the circle TN.

9 As P is a point in the circumference of a semi-circle then the angle between the radius and the tangent is the angle in a semi-circle and must, therefore, be 90°.

Exercises:
10 Set out — full size — the given circle and point P and construct a tangent to the circle to pass through point P.

11 Draw — full size — the given view of a bracket.

12 Set out — full size — the given figure.

If you are not sure how to apply what you have learned to an actual problem, copy the next exercise which has been worked for you.

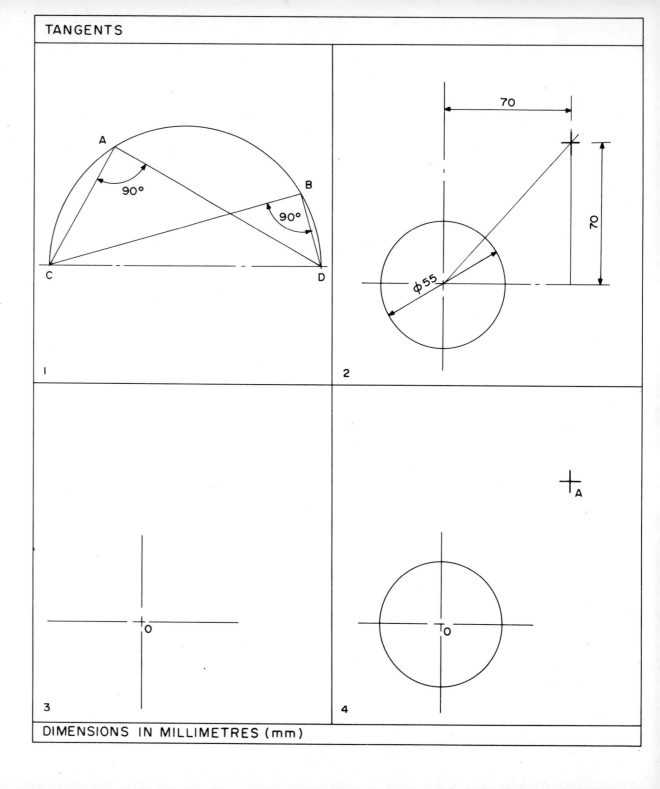

TANGENTS

DIMENSIONS IN MILLIMETRES (mm)

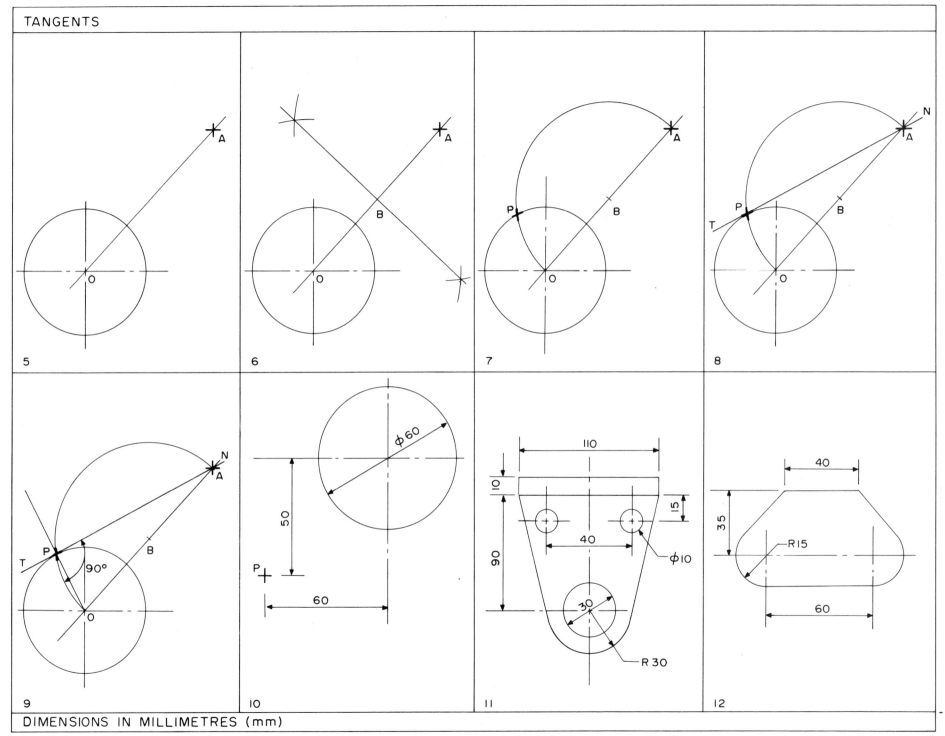

To set out the view of a clip shown in Panel No. 1.
Draw this exercise following, stage by stage, the instructions and the drawings in the panels.

Panel No:

2 Fix the centre of the curves O and the point A.

   With centre O draw a curve of 40 mm radius.

   Join O and A with a straight line and bisect it to fix point B.

3 With the point of your compasses at B and with radius BA draw the semi-circle OA. This fixes point P where the two curves intersect.

4 Join points A and P with a straight line.

5 Join the centre of the curves O and point P with a straight line.

6 With centre at O and compasses set to 30 mm draw the inner curve which fixes point Q on line OP.

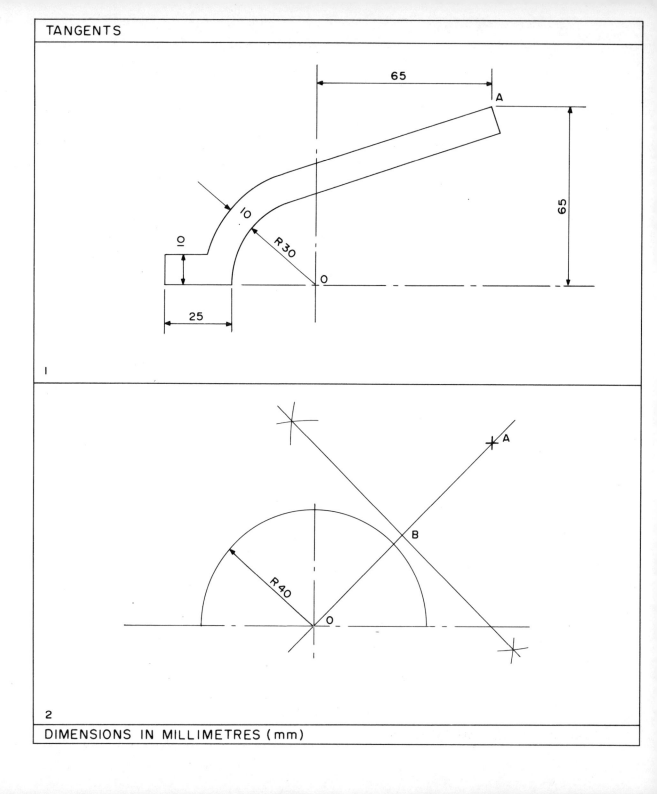

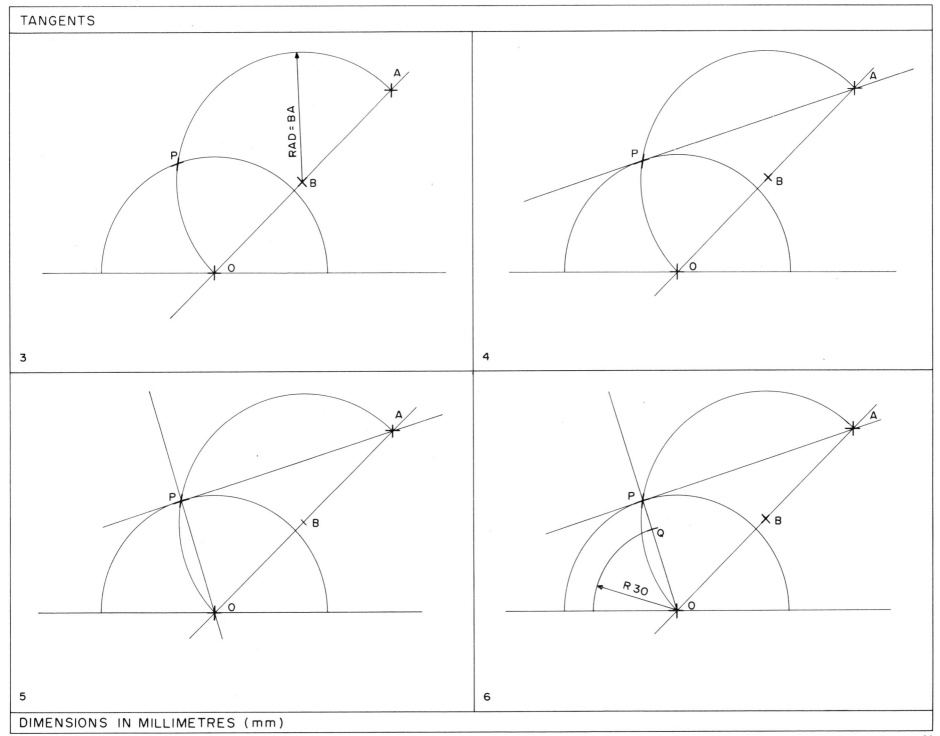

7   Parallel to AP draw straight line QD.

8   Draw the remaining detail.

9   Line in your drawing. Remember to line in curves first.

10 From any given point such as A outside a circle there are, of course, *two* tangents shown here as TN and T^1N^1.

11 A line such as OM drawn at 90° to a tangent and passing through the point where the tangent touches the circumference of the circle is said to be 'Normal' to the curve and is sometimes called a Normal.

    For a line to be normal to a circular arc, it must pass through its centre.

Exercise:
12 Set out the figure full size.

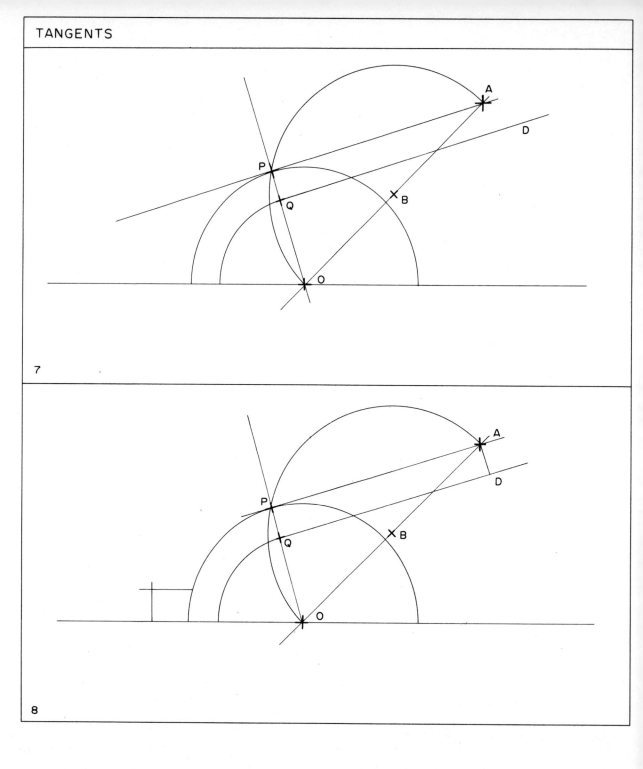

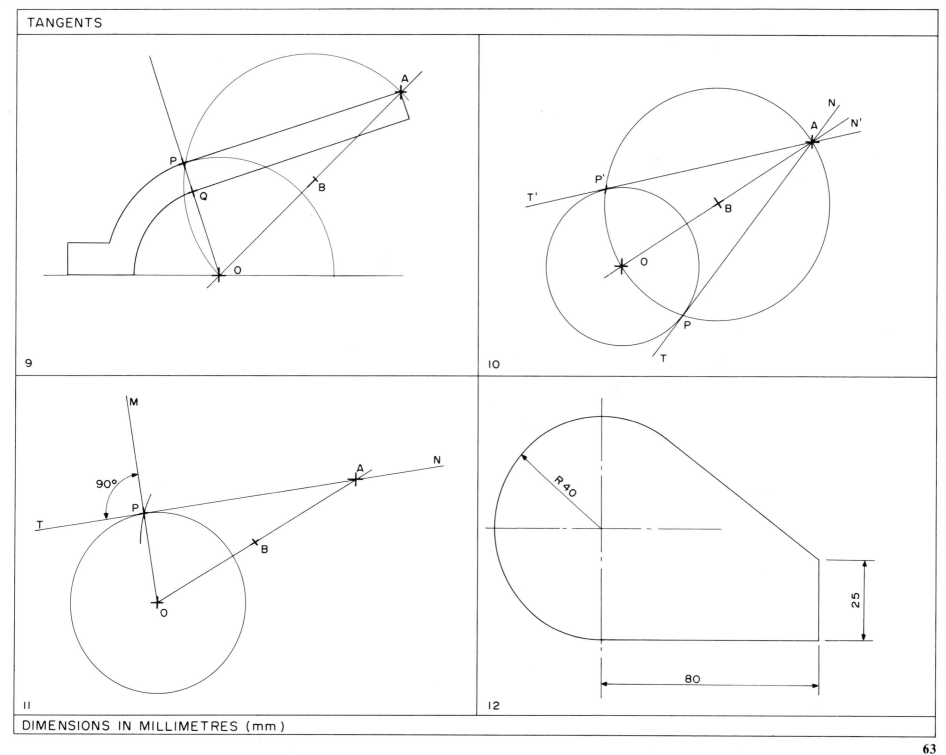

Rounding off corners

In Technical Drawing it is often necessary to set out rounded corners similar to those shown in Panel No. 1. Before learning how to do this there are a number of things you should note.

1. The 'rounding off' should be smooth and straight lines and curves should meet without a 'kink'.

2. The curves, as shown in Panel No. 2, are parts of circles and the straight lines are tangents to them. (This means that they touch the curve at ONE point only).

3. It is important to understand WHERE the straight line and the curve meet and this is shown (for each corner) in Panels Nos. 3, 4, 5 and 6.

   A line drawn through the centre of the curve at 90° to the straight line gives the point where the straight line changes to a curve.

4. When the corner to be rounded off is a right angle (as in Panels Nos. 3 and 5) the curve will be a quarter circle (or quadrant). When the corner is an acute angle (as in Panel No. 4) the curve will be greater than a quarter circle and when the corner is an obtuse angle (as in Panel No. 6) the curve will be less than a quarter circle.

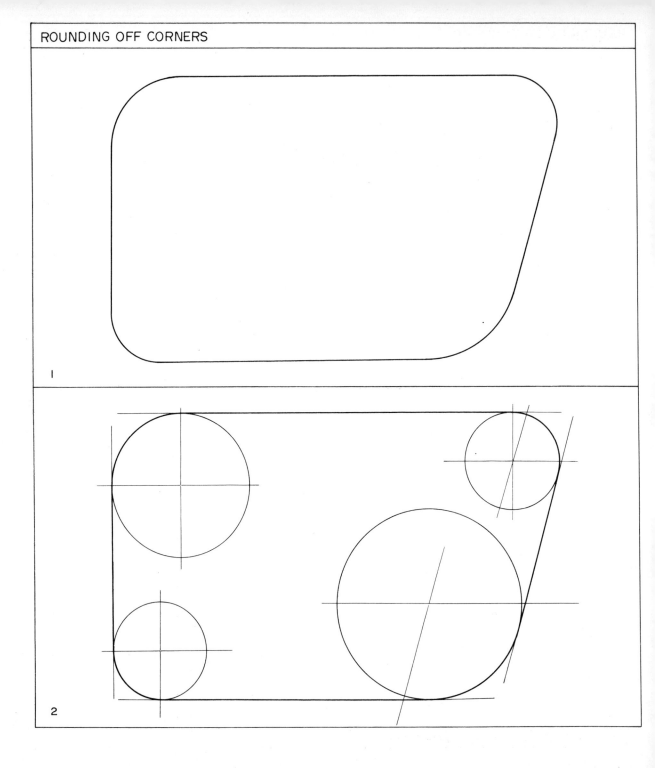

ROUNDING OFF CORNERS

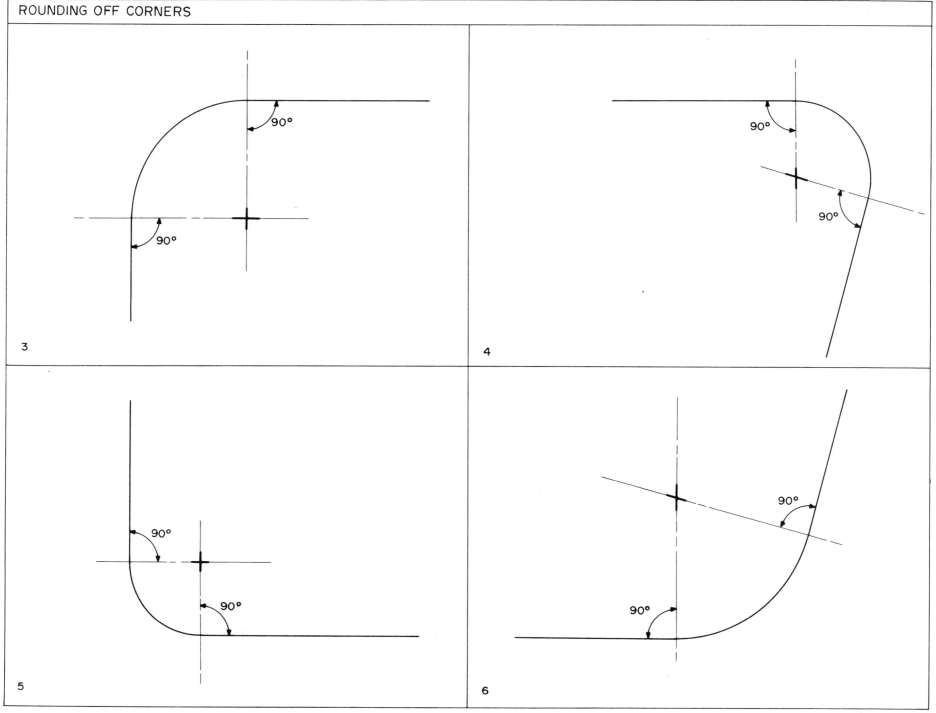

To draw a curve to round off a right-angled corner.
Draw this exercise following, stage by stage, the instructions and the drawings in the panels.

Panel No:
1. This shows a right-angled corner rounded off to a radius of 20 mm. Remember — the straight lines are tangential to the curve — they touch but do not cut it.
2. If every circle of 20 mm radius which touches, and is to the left of, line AB is drawn then their centres will all lie on line EF which is parallel to, and 20 mm away from, line AB. Therefore, the centre of our curve must lie in line EF.
3. Similarly, if every circle of 20 mm radius which touches, and is above, line BC is drawn then their centres will all lie on line GH which is parallel to and 20 mm away from line BC. Therefore, the centre of our curve must also lie on line GH.

   As the centre of our curve lies in both line EF and line GH it must be where they meet (or intersect).
4. Draw the two lines AB and CB at right angles.
5. Draw line EF parallel to and 20 mm away from line AB.
6. Draw line GH parallel to and 20 mm away from line CB.

   This fixes point O which is the centre of the required curve.
7. With compasses set at 20 mm draw the curve from F to H.
8. The full circle has been drawn to show:-
   a) that the required curve is a quadrant,
   b) the lines AB and CB touch but do not cut the curve.

Exercises:
9. Set out the rounded corner.
10. Set out the given figure. The rounded corner is a right angle.
11. Set out the given section.
12. Set out the given figure. The curve at the external corner is 15 mm radius and at the internal corner 10 mm radius.

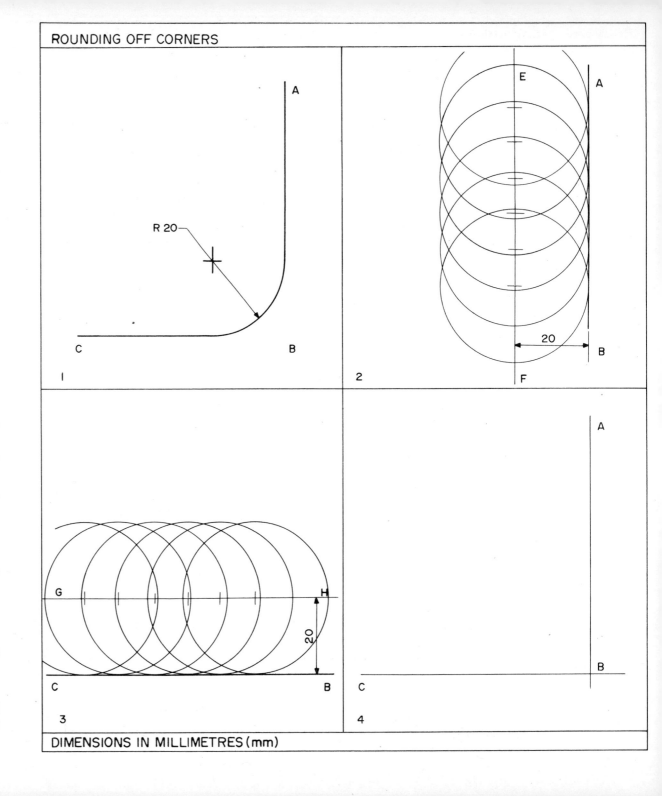

ROUNDING OFF CORNERS

DIMENSIONS IN MILLIMETRES (mm)

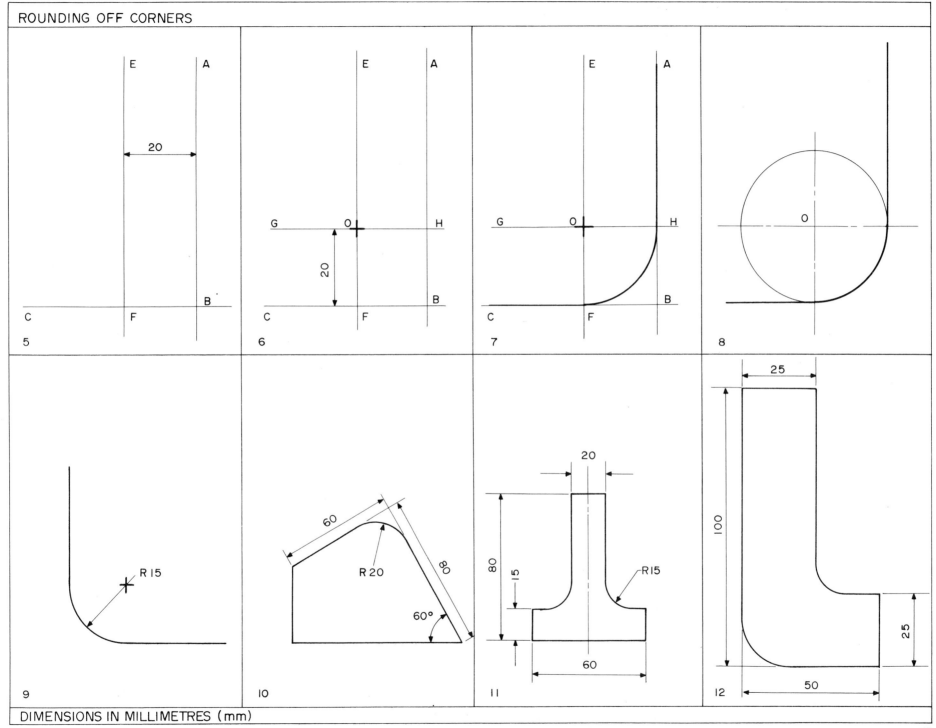

To draw a curve to round off an obtuse-angled corner.

Draw this exercise following, stage by stage, the instructions and the drawings in the panels.

Panel No:
1. This shows an obtuse-angled corner rounded off to a radius of 40 mm. Once again the lines are tangential to the curve — they touch but do not cut it.
2. Draw the two lines AB and BC with an angle of 120° between them.
3. Draw line EF parallel to and 40 mm away from line AB. Remember to hold your ruler at 90° to line AB when measuring the 40 mm.
4. Draw line GH parallel to and 40 mm away from line BC.

   This fixes point O which is the centre of the required curve.
5. From O draw lines OT and OT^1 at 90° to lines AB and BC respectively.
6. With compasses set at 40 mm the curve can now be drawn from T to T^1.

   The full circle has been drawn to show:-

   a) that the required curve is less than a quadrant,

   b) that the lines AB and BC touch but do not cut the curve.

Exercises:
7. Set out full size the rounded corner.
8. Draw, full size, the given tee section.
9. Draw the given bracket full size.
10. Draw the given view of a swan-necked branch pipe. Scale one-fifth size.

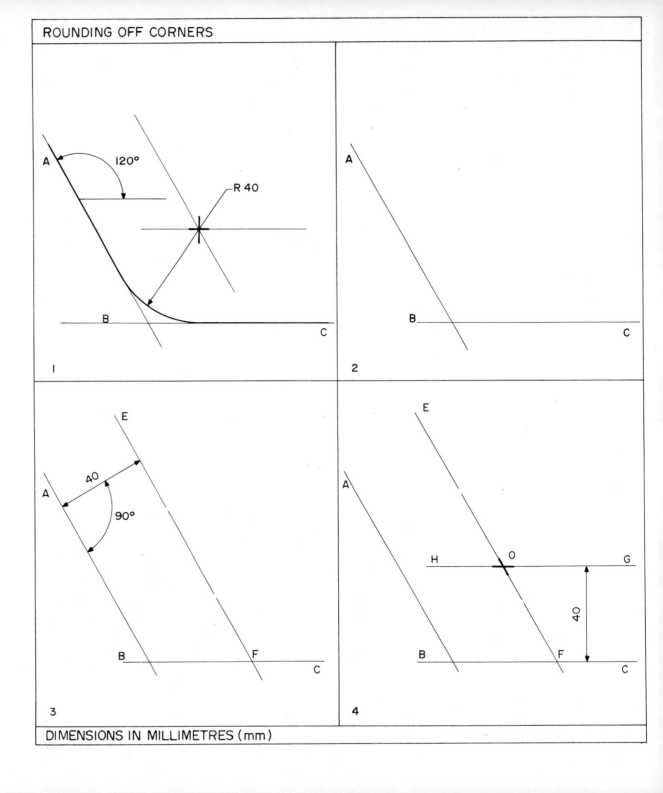

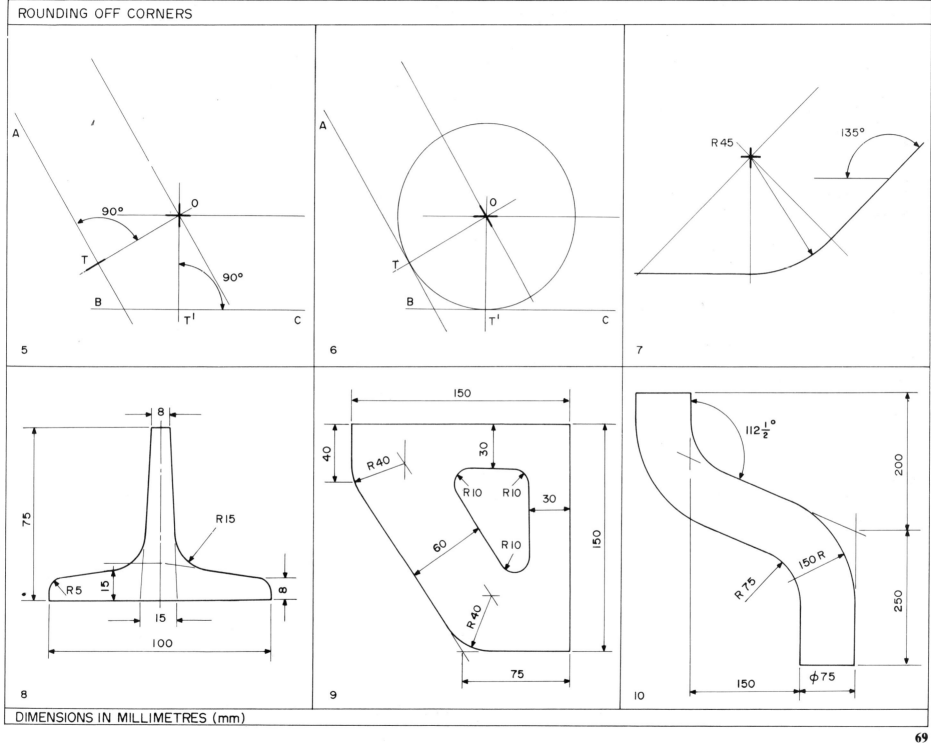

To draw a curve to round off an acute-angled corner.
Draw this exercise following, stage by stage, the instructions and the drawings in the panels.

Panel No:
1. This shows an acute-angled corner rounded off to a radius of 30 mm. Remember — the straight lines are tangential to the curve — they touch but do not cut it.

2. Draw the two straight lines AB and BC with an angle of 60° between them.

3. Draw line EF parallel to and 30 mm away from AB. Be sure to hold your ruler at 90° to line AB when measuring the 30 mm.

4. Draw line GH parallel to and 30 mm away from BC.

   This fixes point O which is the centre of the required curve.

5. From O draw lines OT and OT^1 at 90° to lines AB and BC respectively.

6. With compasses set at 30 mm the curve can now be drawn from T to T^1.

   The complete circle has been drawn to show:-
   a) that the required curve is greater than a quadrant,
   b) that the lines AB and BC are tangential to the curve.

Exercises:
7. Set out the rounded corner.

8-10 Set out the given figures — scale full size.

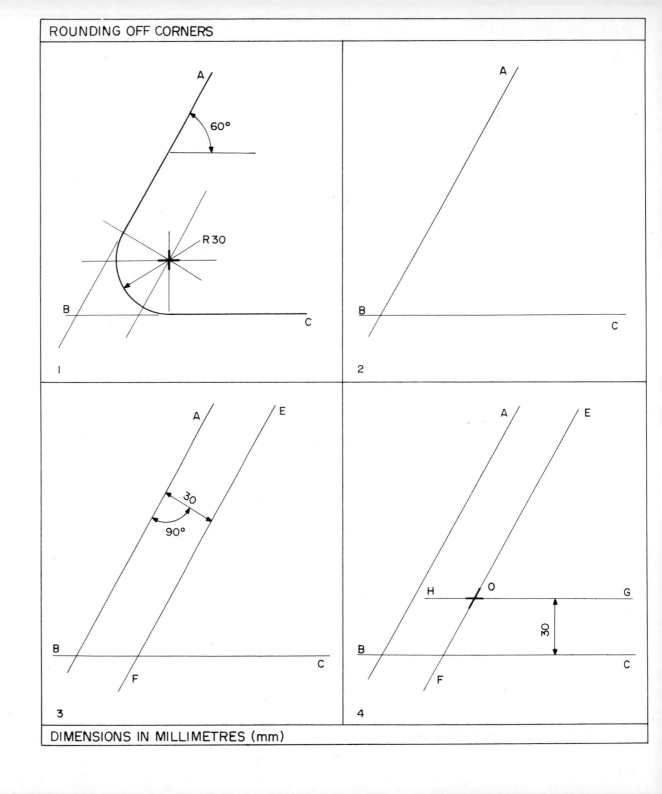

70

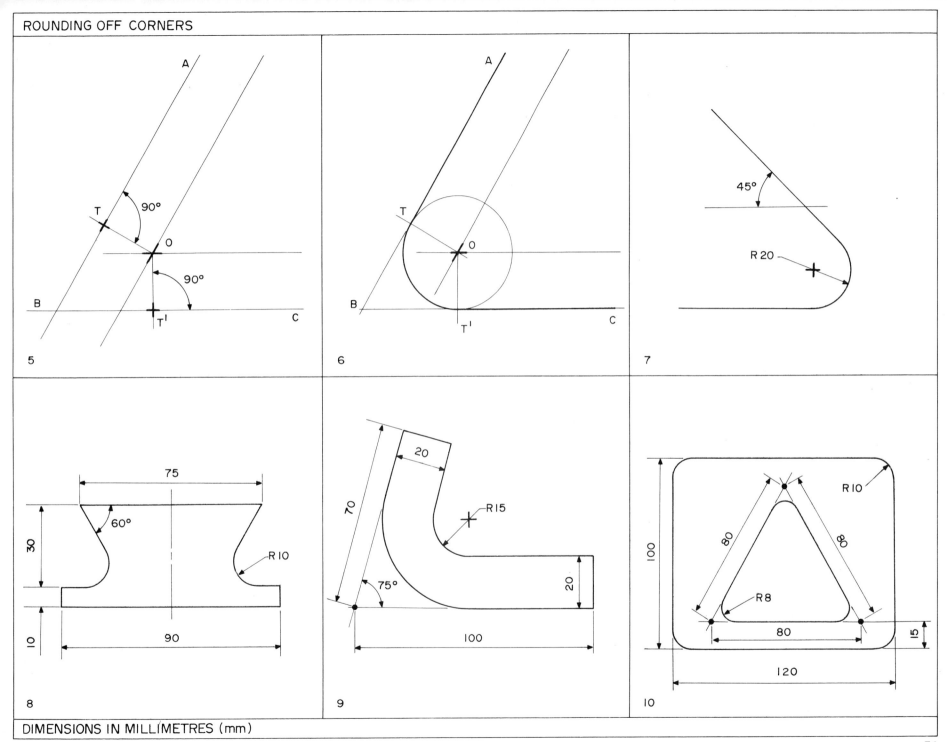

Circle to touch a given circle and straight line

To set out the figure shown in Panel No. 1.

Draw this exercise following, stage by stage, the instructions and the drawings in the panels.

Panel No:

1. This shows a 35 mm radius curve and a straight line joined by a 20 mm radius curve. Notice that where curve meets curve or straight line there are no 'kinks'.

2. This shows the curves as parts of circles. Note the point P where the two curves meet and point P^1 where the straight line and curve meet.

3. If every circle of 20 mm radius which touches the line AB is drawn then their centres will all lie in the line EF which is parallel to and 20 mm away from line AB.

4. If every circle of 20 mm radius which touches the arc CD is drawn then their centres lie in the arc GH which is concentric to CD and 20 mm away from it.

5. Draw centre lines to fix point O and with your compasses set at 35 mm draw the arc CD.

6. Draw line AB 15 mm below centre O.

7. Draw line EF parallel to and 20 mm away from line AB.

8. With your compasses set at 55 mm (35 mm plus 20 mm) from centre O draw the arc GH.

   Where the arc GH cuts the line EF we have O^1 which is the centre of the 20 mm radius curve.

9. Join centres O^1 and O with a straight line to fix point P where the change of curve occurs. From O^1 draw a line at 90° to line AB to fix point P^1 where the straight line meets the curve.

10. With your compasses set to 20 mm from centre O^1 draw a curve from P^1 to P.

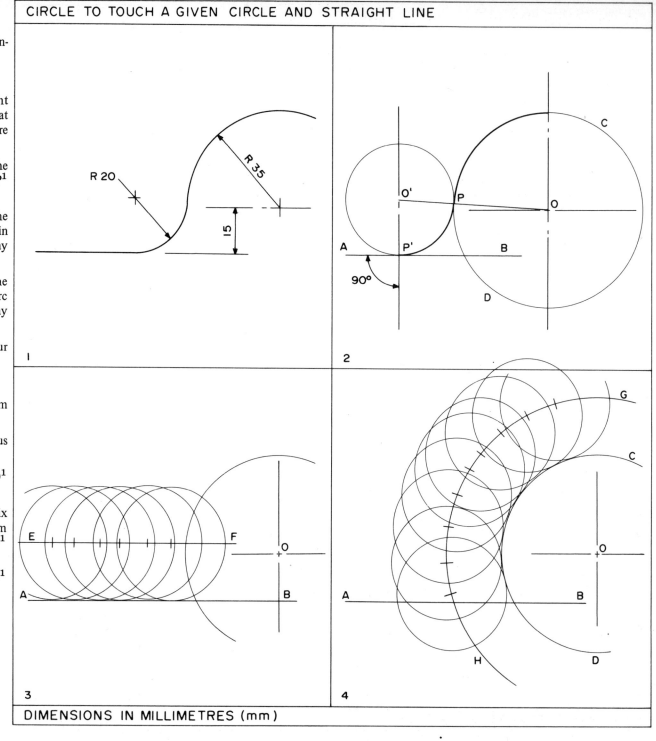

DIMENSIONS IN MILLIMETRES (mm)

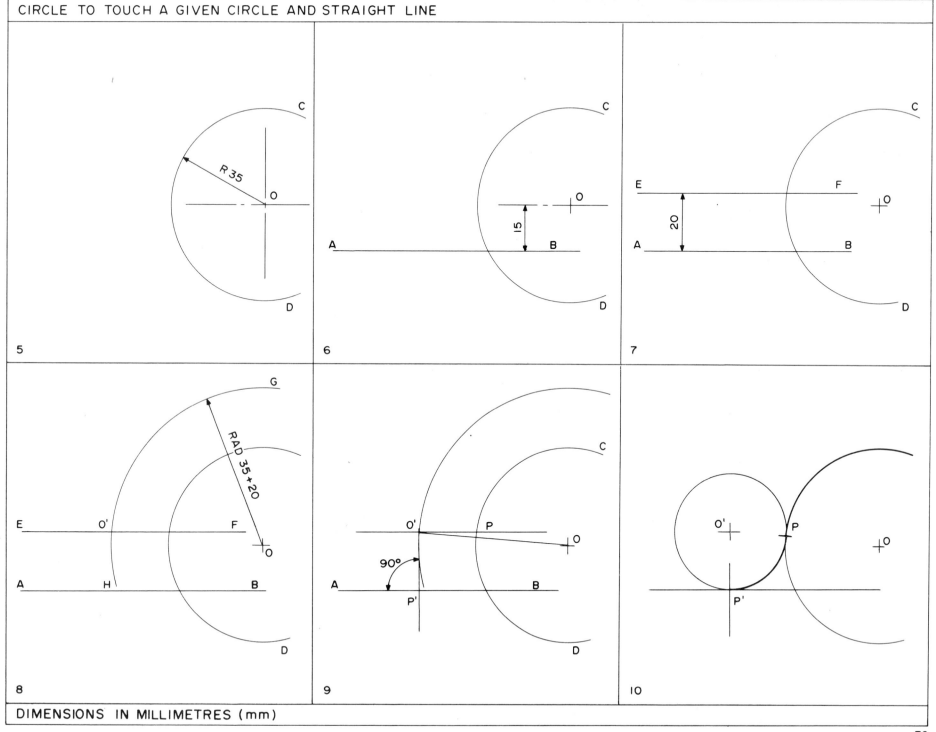

To draw the view of a clock case shown in Panel No. 1.

Draw this exercise following, stage by stage, the instructions and the drawings in the panels.

Panel No:

1. This shows the elevation of a clock case. Note that it is symmetrical.

2. Fix centre O and with your compasses set to 81 mm draw arc CD.

3. 25 mm below centre O draw lines AB and A^1B^1.

4. The drawings in the remainder of the panels are to a larger scale and only the left hand half of the drawing is shown. As the drawing is symmetrical the construction for the right hand side will be the same.

   Draw line EF parallel to and 50 mm above line AB.

5. With your compasses set to 131 mm (81 mm plus 50 mm) draw arc GH.

   The intersection of arc GH and straight line EF fixes centre O^1.

6. Join centres O^1 and O with a straight line to fix point P where the change of curve occurs.

   From O^1 draw a straight line at 90° to line AB to fix point P^1 where the straight line meets the curve.

7. With your compasses set to 50 mm draw the curve from P to P^1.

8. Now draw the 150 mm diameter circle and complete and line in the drawing.

Exercises:

9. Draw the given figure – full size.

10. Draw the keyhole cover – scale five times full size.

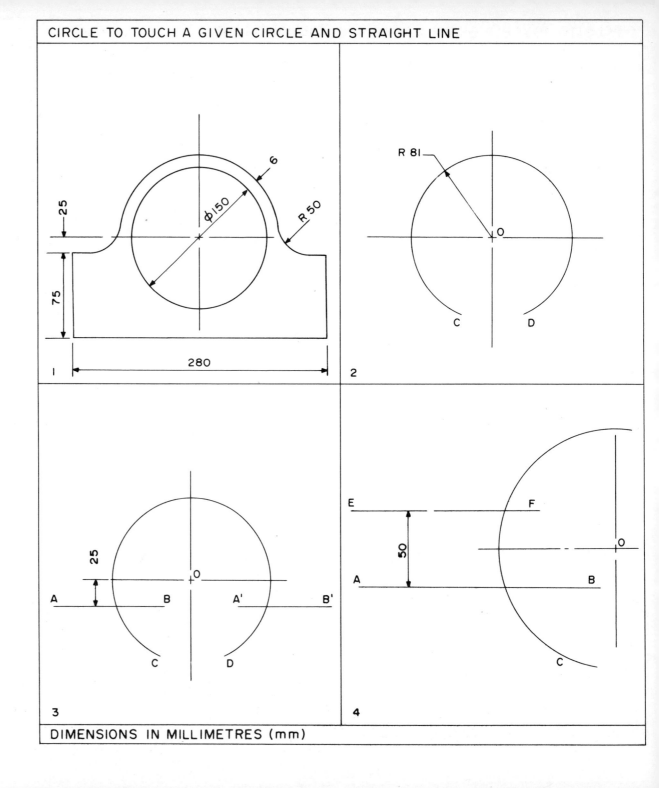

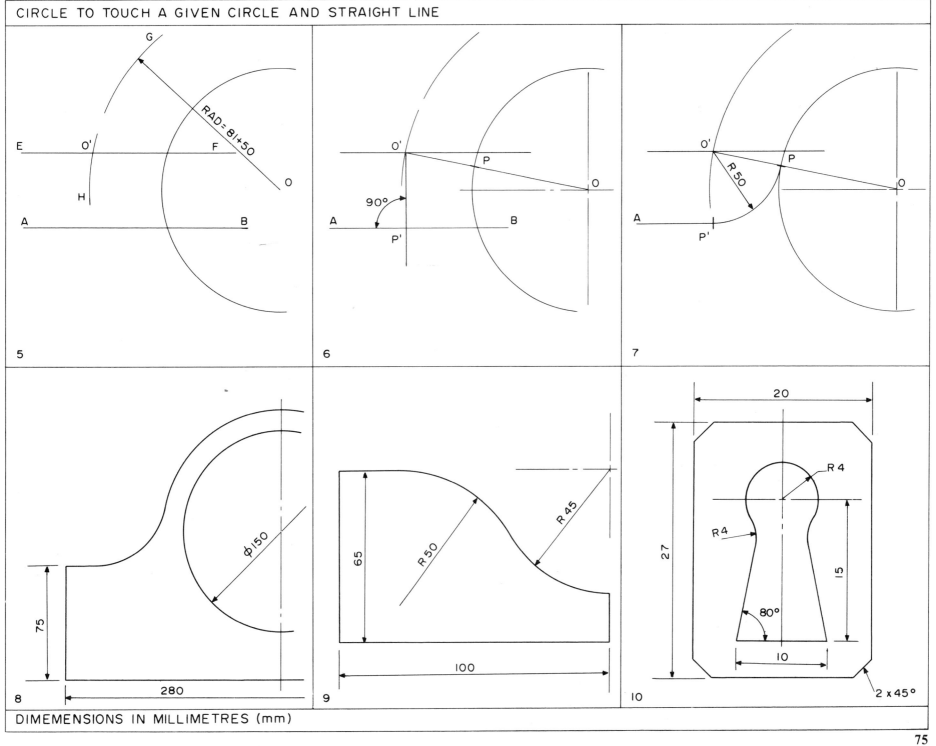

Circles in contact

To draw the figure shown in Panel No. 1.
Draw this exercise following, stage by stage, the instructions and the drawings in the panels.

Panel No:

1. This shows a figure consisting of three separate arcs. The positions of the centres of two of them are given but the centre for the third must be found.

2. This shows the circles of which the arcs are a part.

3. Draw a centre line at 60° and fix centres O and O^1 100 mm apart.

4. With centre O draw a 50 mm radius circle and with centre O^1 a 30 mm radius circle.

5. With your compasses set to 90 mm (50 mm plus 40 mm) from centre O draw the arc EF.

6. With your compasses set to 70 mm (30 mm plus 40 mm) from centre O^1 draw the arc GH.

   The intersection of arcs EF and GH fixes centre O^2.

7. Join O^2 and O and O^2 and O^1 with straight lines to fix points P and P^1 where the changes of curve occur.

8. With your compasses set to 40 mm from centre O^2 draw the curve from P to P^1.
   (Note: If curves are to be joined without 'kinks' then a line joining their centres always passes through the point at which the change of curve occurs).

Exercises:

9. Set out the given figure – full size.

10. Set out the section of moulding – full size.

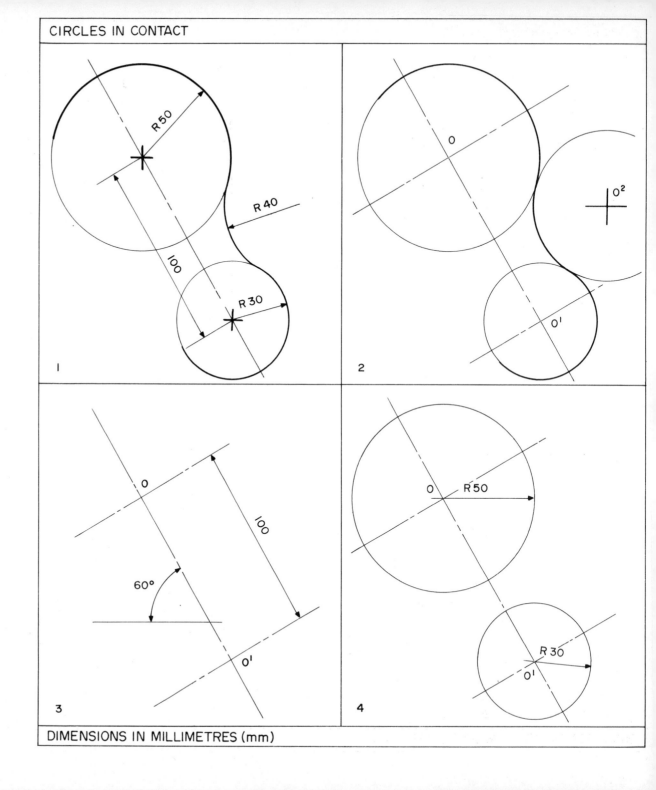

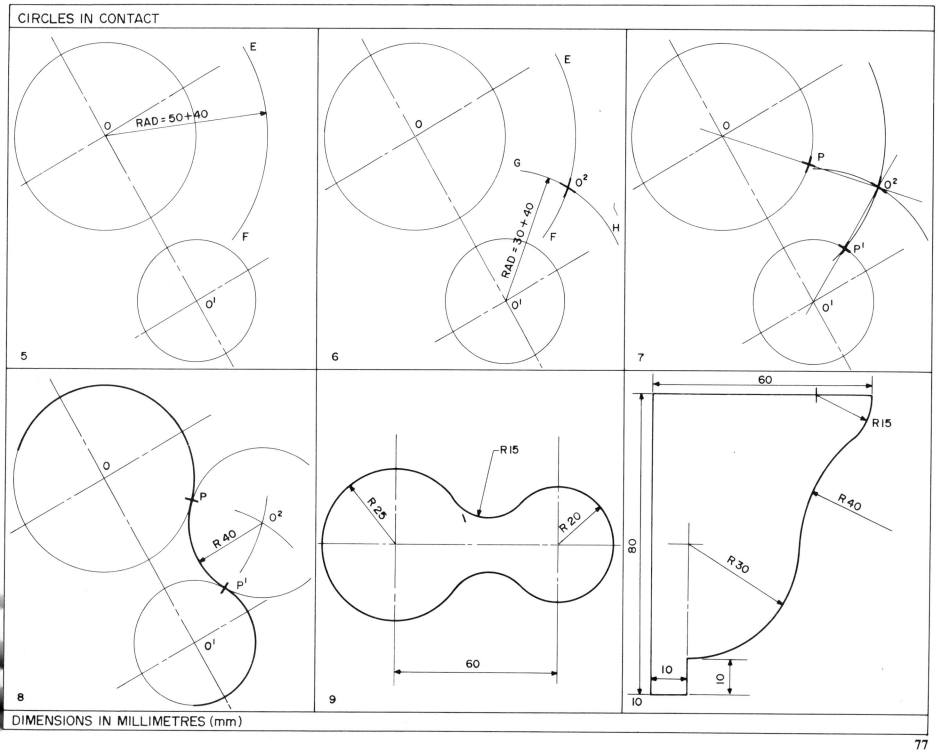

To draw the figure shown in Panel No. 1.
Draw this exercise following, stage by stage, the instructions and the drawings in the panels.

Panel No:

1. In the figure shown all necessary dimensions are given but the position of the centre for the 100 mm radius curve will have to be found geometrically.

2. This panel shows that what we need to do it to draw a circle to touch and enclose two given circles.

   If you look back to the previous exercise you will see that there we were drawing a circle to touch but exclude two given circles.

3. Draw the centre line at 60° and fix centres O and O^1 85 mm apart.

4. With your compasses set to 30 mm, from centre O draw arc AB.

   With your compasses set to 40 mm, from centre O^1 draw arc CD.

5. With your compasses set to 70 mm (100 mm minus 30 mm) from centre O draw arc EF.

6. With your compasses set to 60 mm (100 mm minus 40 mm) from centre O^1 draw arc GH.

   The intersection of the arcs EF and GH fixes centre O^2.

7. Join centres O^2 and O and O^2 and O^1 with straight lines to fix points P and P^1 where changes of curve occur.

8. With your compasses set to 100 mm from centre O^2 draw an arc from P to P^1.

9. The drawing can now be completed.

Exercise:

10. Draw, full size, the figure shown.

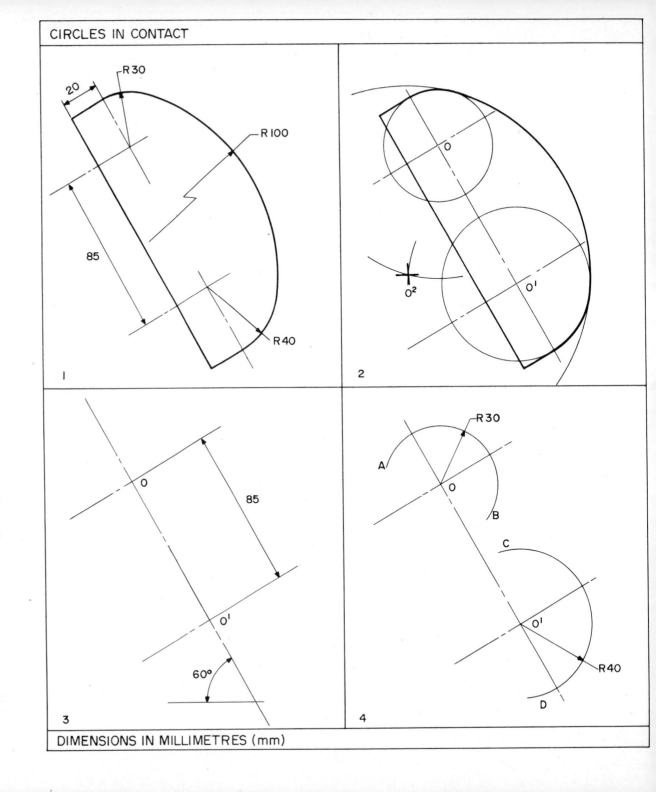

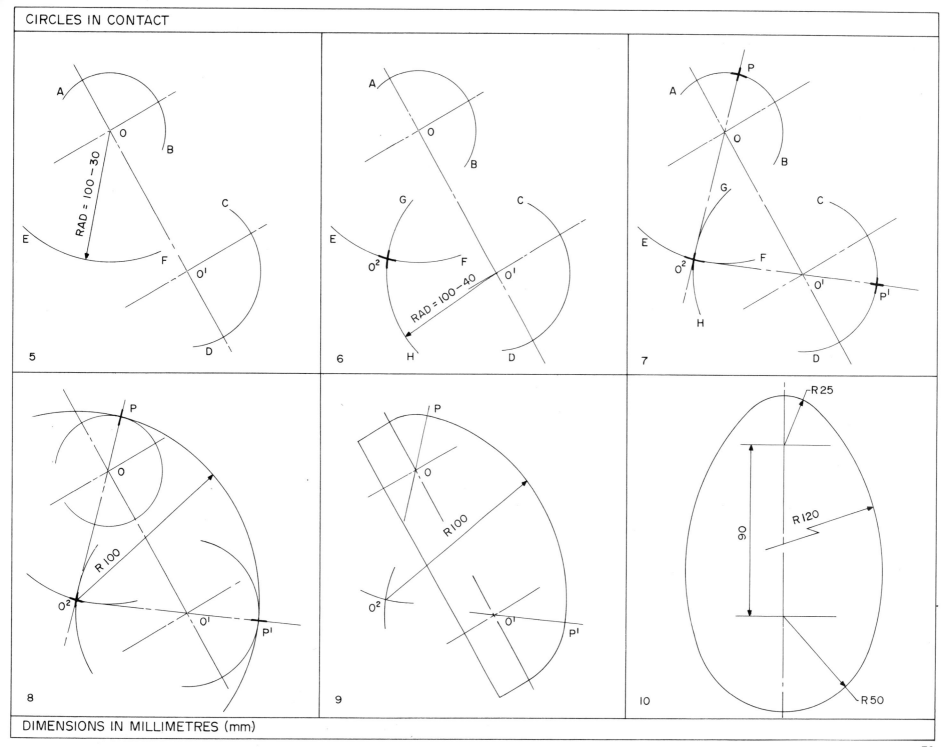

To set out the figure shown in Panel No. 1.
Draw this exercise following, stage by stage, the instructions and the drawings in the panels.

Panels No:

1. This drawing gives all the necessary information the problem being to find, geometrically, the centre of the 70 mm radius curve.

2. This shows the curves as parts of circles. The positions of the circle centre O and the point A can be fixed, the problem being to draw a circle to pass through a given point A and to touch and exclude a given circle.

   Note in the last exercise the problem was to draw a circle to touch and include two given circles.

3. Draw centre lines and with compasses set at 45 mm draw circle centre O.

4. Fix the position of point A.

5. With compasses set to 115 mm (45 + 70 mm), and the point of the compasses at O, draw the arc EF.

6. With compasses set to 70 mm and with Point A as the centre draw arc GH. The intersection of the Arcs EF and GH fixes O^1 the centre for the 70 mm radius curve.

7. A straight line drawn from O^1 through O fixes point P where the change of curve occurs.

8. With compasses set at 70 mm and with centre O^1 the curve can be drawn from P to A.

9. This shows a circle centre O drawn to touch but exclude a given circle centre O^1 and passing through a given point A.

   Note point P where the circles touch lies on a line joining their centres.

Exercise:

10. Draw this figure full size.

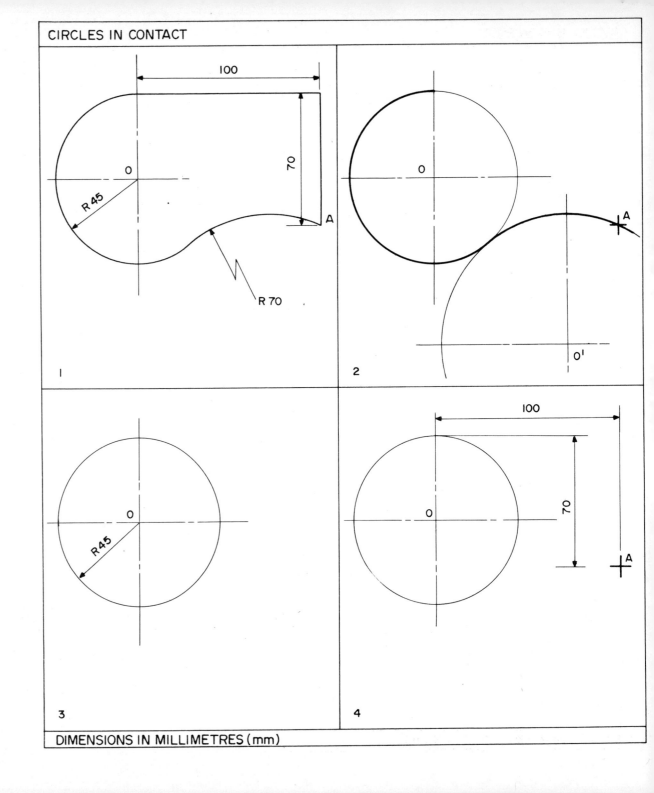

80

CIRCLES IN CONTACT

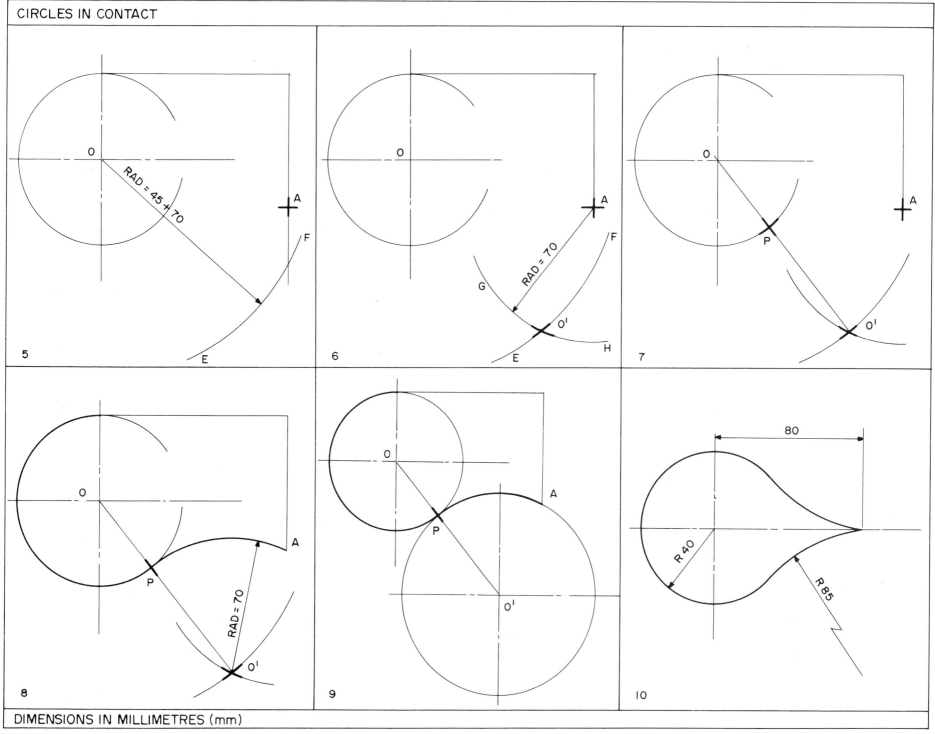

DIMENSIONS IN MILLIMETRES (mm)

To set out the figure in Panel No. 1.
Draw this exercise following, stage by stage, the instructions and the drawings in the panels.

Panel No:

1. This drawing gives all the necessary information, the problem being to find, geometrically, the centre of the 85 mm radius curve.

2. This shows the curves as parts of circles. The positions of the circle centre O and the point A can be fixed, the problem being to draw a circle to pass through a given point A and to touch and include a given circle.

3. Draw centre lines and with compasses set at 35 mm draw circle centre O.

4. Fix the position of point A.

5. With compasses set at 50 mm (85 mm minus 35 mm) and the point of the compasses at O draw the arc EF.

6. With the compasses set at 85 mm and with point A as the centre draw the arc GH. The intersection of the arcs EF and GH fixes O^1 the centre for the 85 mm curve.

7. A straight line drawn from O^1 through O fixes point P where the change of curve occurs.

8. With your compasses set at 85 mm and with centre O^1 the curve can be drawn from P to A.

Exercises:

9, 10   Draw each of these figures full size.

Show clearly your construction for finding centres of curves.

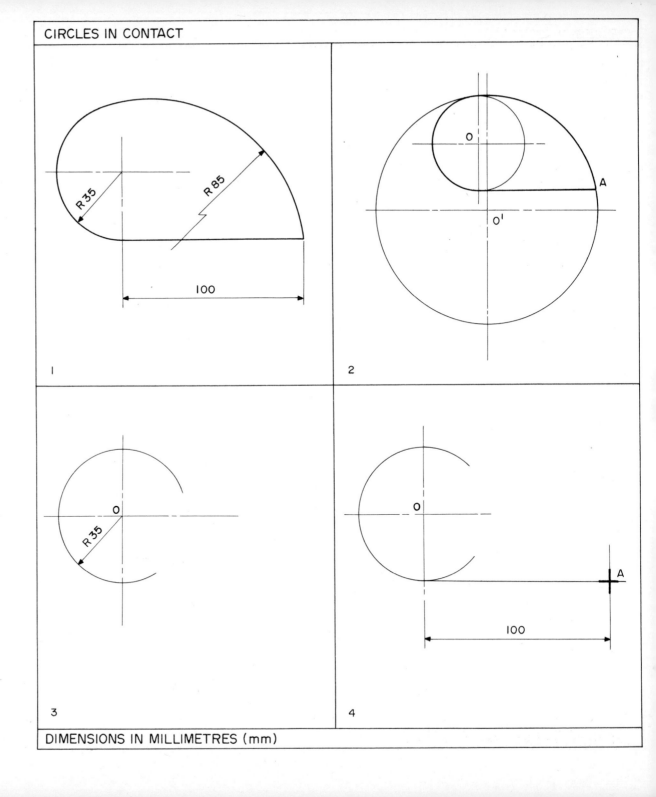

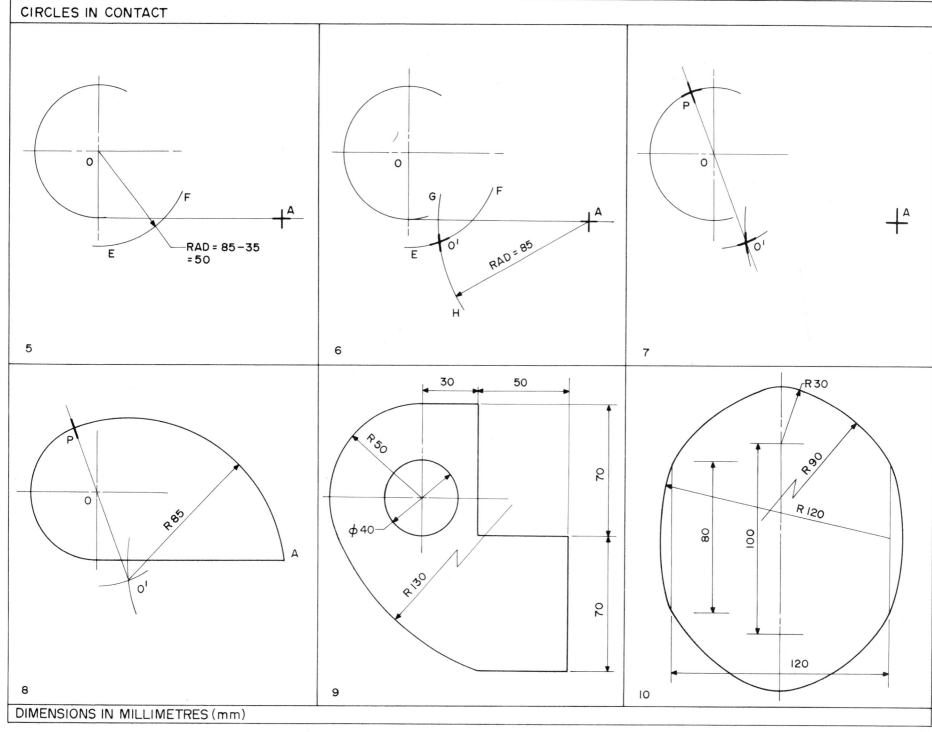

To set out the figure in Panel No. 1.
Draw this exercise following, stage by stage, the instructions and the drawings in the panels.

Panel No:

1. This drawing gives all the necessary information the problem being to find, geometrically, the centres of the 20 mm radius curves.

2. This shows the curves as parts of circles which touch a given straight line and a given curve.

3. Draw centre lines and with your compasses set at 100 mm draw the curve centre O.

4. Measure 30 mm either side of O to fix points C and C^1 and draw the straight lines CD and C^1D^1 at 75° to the horizontal.

5. Draw lines EF and E^1F^1 20 mm away from and parallel to CD and C^1D^1 respectively.

6. With your compasses set at 80 mm (100 - 20) draw the arc GH. The intersections of the arc GH with lines EF and E^1F^1 fixes the centres of the 20 mm radius curve O^1 and O^2.

7. This is an enlarged view of one side and shows how to fix point P where a change of curve occurs and point P^1 where the 20 mm curve meets the straight line.

    A straight line drawn through O and O^2 fixes point P.

    A straight line drawn through O^2 at 90° to C^1D^1 fixes point P.

8. This is an enlarged view showing the 20 mm radius curve drawn from P to P^1.

9. Line in your drawing.

Exercise:
10. Draw this figure full size.

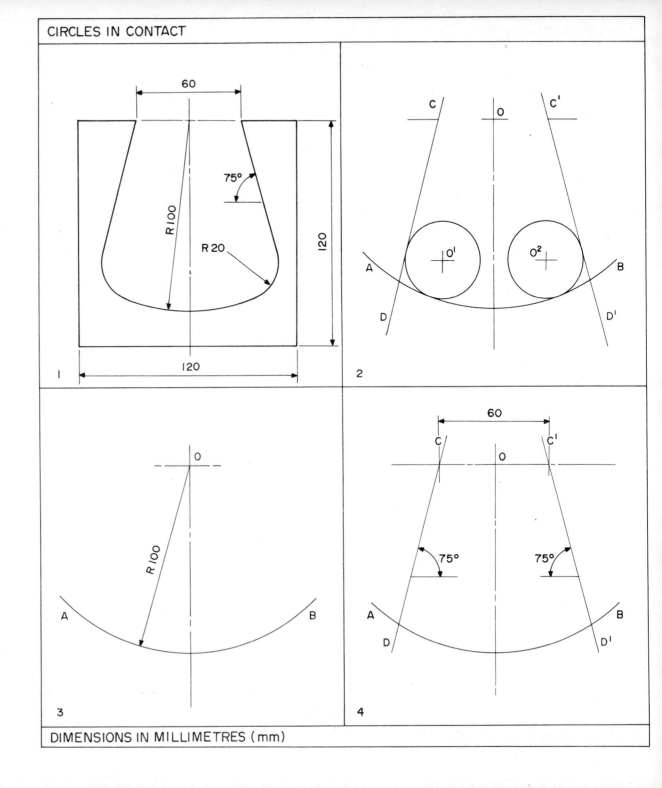

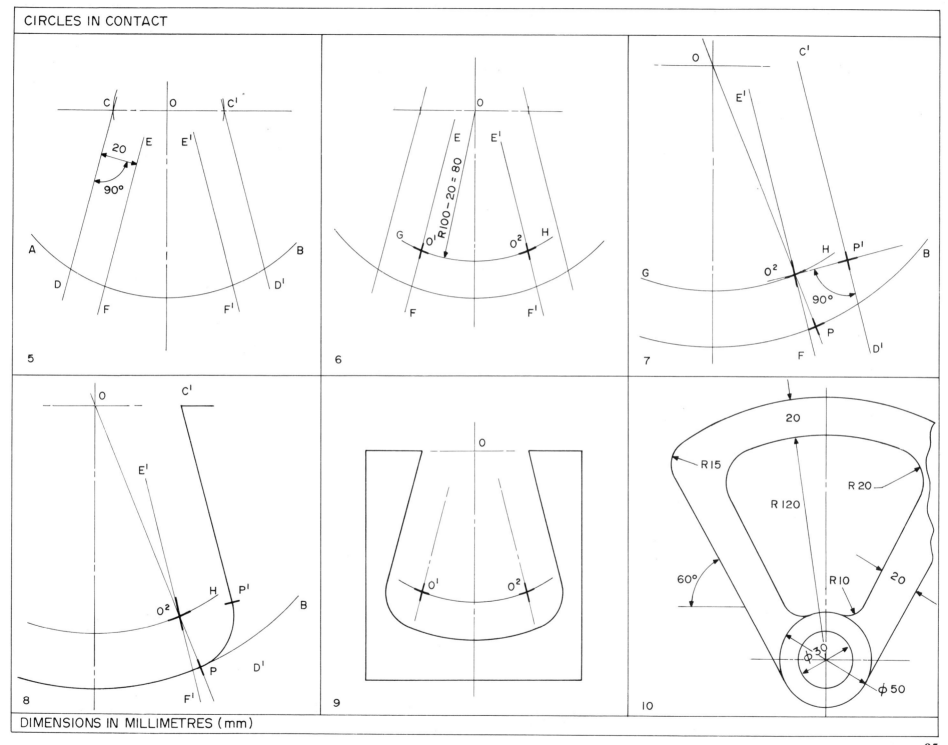

To set out the figure in Panel No. 1.
Draw this exercise following, stage by stage, the instructions and the drawings in the panels.

Panel No:

1. This drawing gives all the necessary information, the problem being to find, geometrically, the centre of the 70 mm radius curve.

2. This shows the 70 mm radius curve as part of a circle which touches and excludes one circle and touches and includes another.

3. Fix the centres O and O^1.

4. With your compasses set at 35 mm draw a circle with centre O^1.

   With your compasses set at 15 mm draw a circle with centre O.

5. With your compasses set at 55 mm (70 - 15) and the point of your compasses at O, draw the arc EF.

6. With your compasses set at 105 mm (70 + 35) and the point of your compasses at O^1 draw the arc GH.

   The intersection of the arcs EF and GH fixes O^2 the centre of the 70 mm radius curve.

7. A straight line drawn through O and O^2 fixes point P and a straight line drawn through O^1 and O^2 fixes point P^1.

   Points P and P^1 are where the changes of curve occur.

8. With your compasses set at 70 mm and with O^2 as the centre the 70 mm radius curve can now be drawn to complete the drawing.

Exercises:

9. Draw a 120 mm diameter circle to touch and exclude the larger circle and touch and include the smaller.

10. Set out the section of moulding full size.

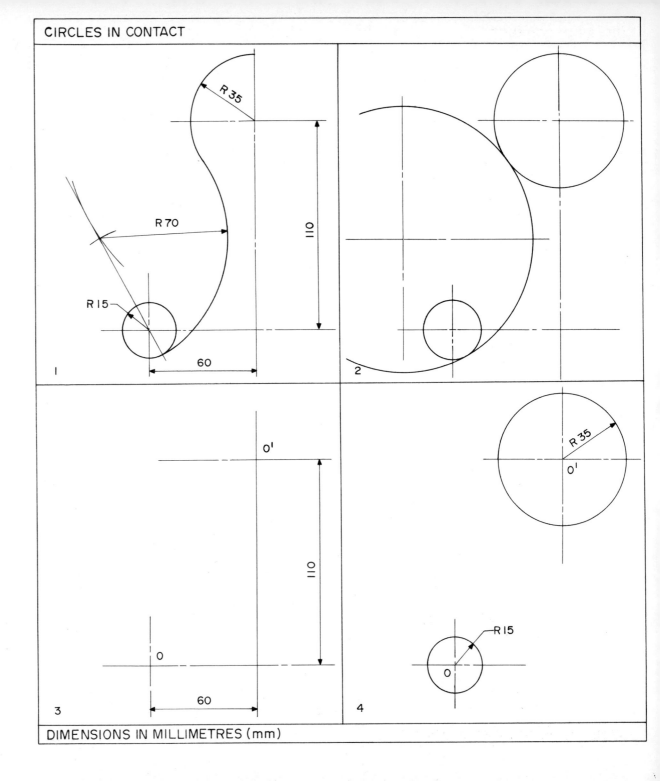

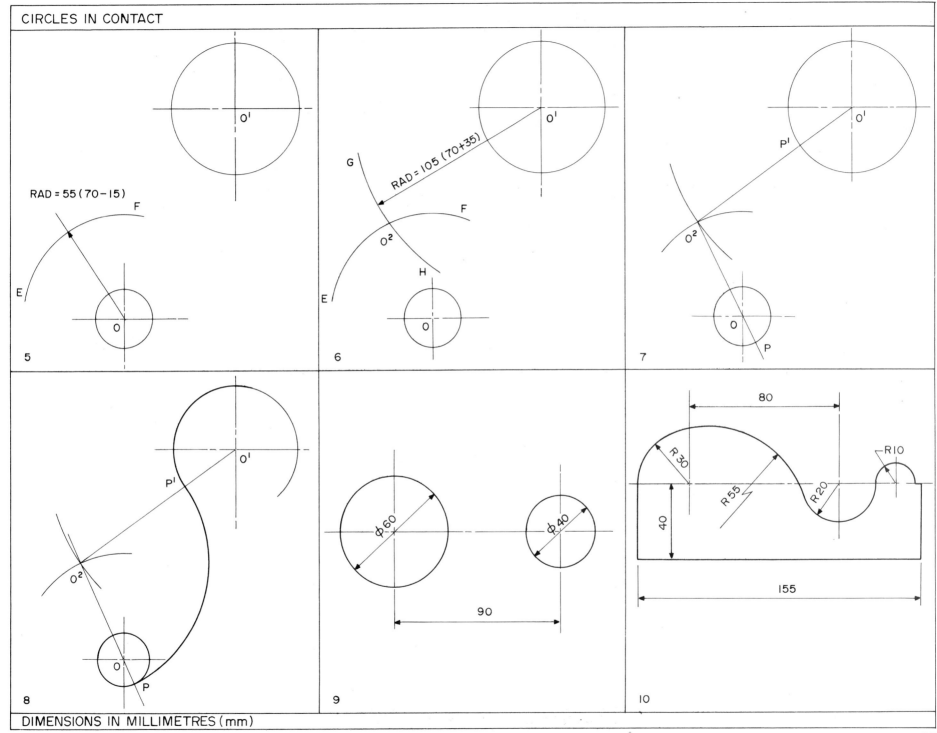

COMMON TANGENTS

We have already dealt with tangents to a single circle. When a straight line is a tangent to two circles it is called a common tangent.

The drawing in Panel No. 1 shows two circles and their two EXTERNAL common tangents TN and T^1N^1.

The drawing in Panel No. 2 shows the same two circles with their two INTERNAL common tangents TN and T^1N^1.

To draw a common external tangent to two given circles.

Draw this exercise following, stage by stage, the instructions and the drawings in the panels.

Panel No:

3 Set out points A and B.

4 With your compasses set at 40 mm and with A as the centre draw a circle.

   With your compasses set at 22 mm and with B as the centre draw a second circle.

5 With your compasses still set at 22 mm place their point at D and mark point E.

6 Place the point of your compasses at A, open them out to point E and draw a circle.

7 Bisect the distance between Centre A and Centre B giving point X. With the point of your compasses at X draw a semi-circle from A to B.

   The intersection of the semi-circle with the circle AE fixes point F.

8 Draw a straight line passing through A and F this fixes point T on the circumference of the 40 mm radius circle.

9 Parallel to AT draw a straight line passing through centre B.

   This fixes point N on the circumference of the 22 mm radius circle.

10 Draw the common external tangent TN.

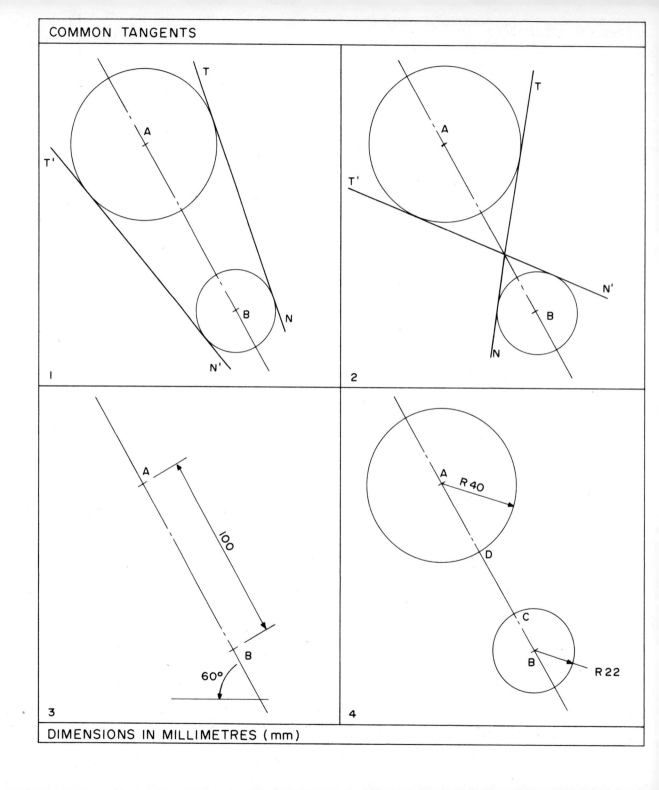

DIMENSIONS IN MILLIMETRES (mm)

COMMON TANGENTS

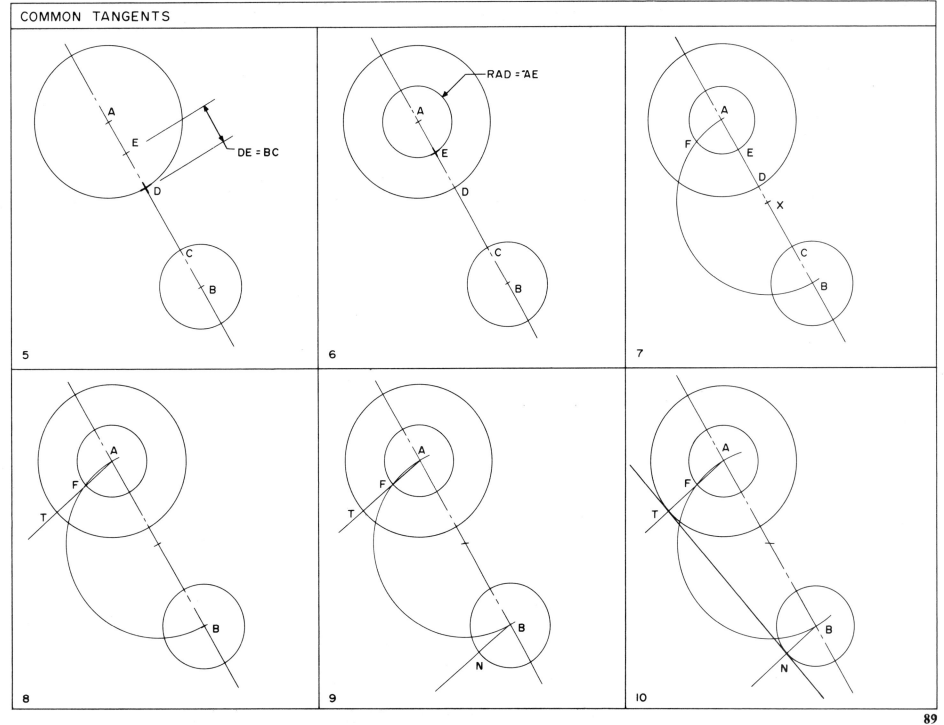

To draw a common internal tangent to two given circles.

Draw this exercise following, stage by stage, the instructions and the drawings in the panels.

Panel No:

1. This shows a common internal tangent.
2. Set out the two circles as shown.
3. Mark point D and C.
4. With your compasses set to 22 mm (the radius of the smaller circle) and with the point of the compasses at D mark point E.
5. Place the point of your compasses at centre A, open them out to point E and draw the arc GH.
6. Bisect the distance between Centre A and Centre B giving point X. With X as the centre draw a semicircle from A to B.

    The intersection of the semi-circle with the arc GH fixes point F.
7. Draw a straight line passing through A and F. This fixes point T in the circumference of the 40 mm radius circle.
8. Parallel to AT draw a straight line through Centre B. This fixes point N in the circumference of the 22 mm radius circle.
9. Draw the common internal tangent TN.

Exercise:

10. Draw the four common tangents to the two given circles.

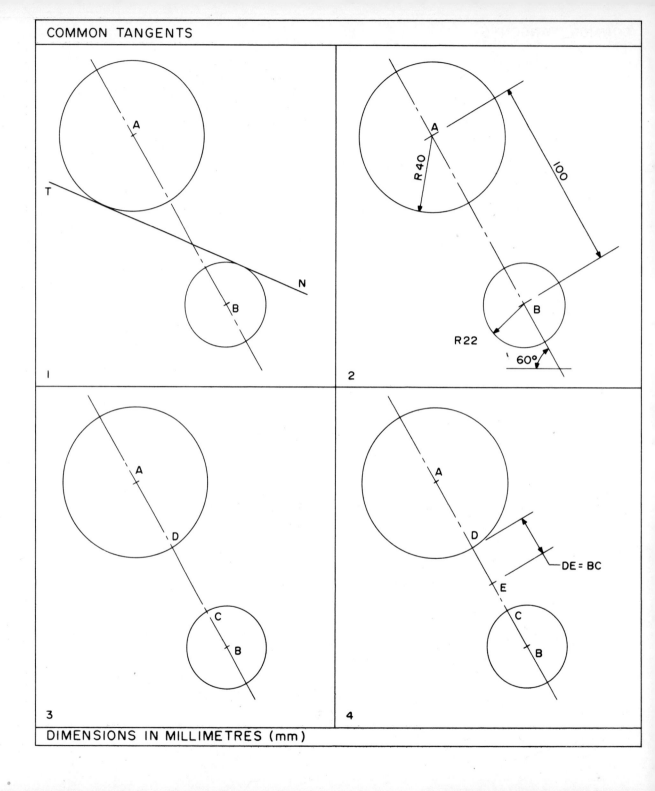

COMMON TANGENTS

DIMENSIONS IN MILLIMETRES (mm)

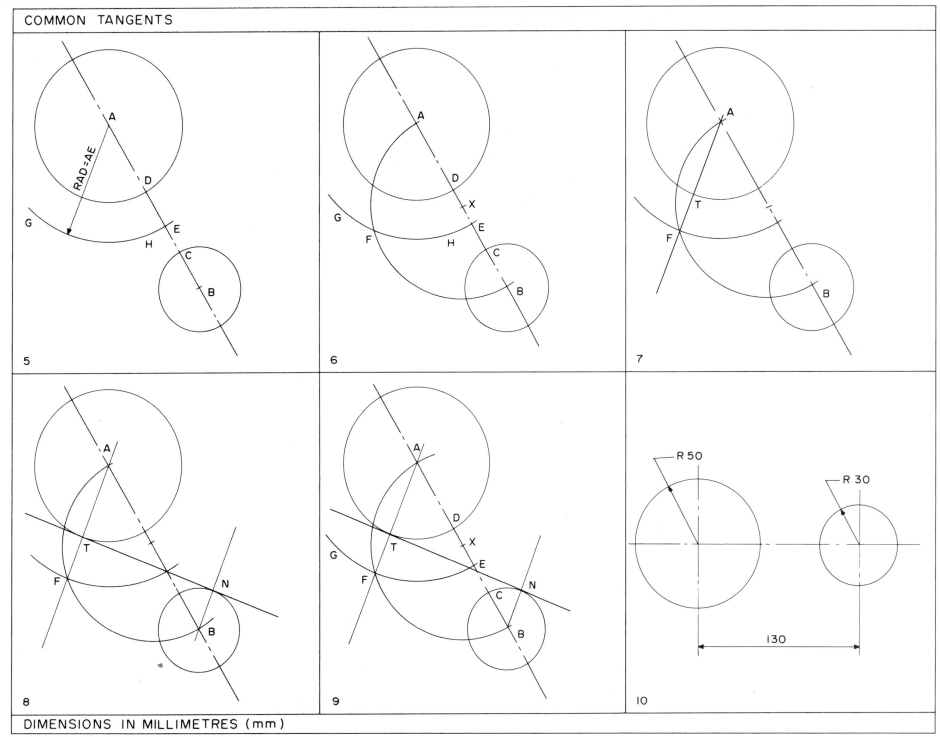

CIRCLES AND TRIANGLES

Panel No:

1. This shows the inscribed circle to a triangle.

2. This shows a triangle with its circumscribing circle.

3. To find the centre for an inscribed circle (as in Panel No. 1) bisect any two angles. The intersection of the bisectors fixes the centre O.

4. To find the centre for a circumscribed circle (as in Panel No. 2) bisect any two sides. The intersection of the bisectors fixes the centre O.

5. This shows the three escribed circles to a triangle.

   Careful study of the drawing should enable you to understand the construction.

Exercise:

Draw any triangle and its inscribed, circumscribed and escribed circles.

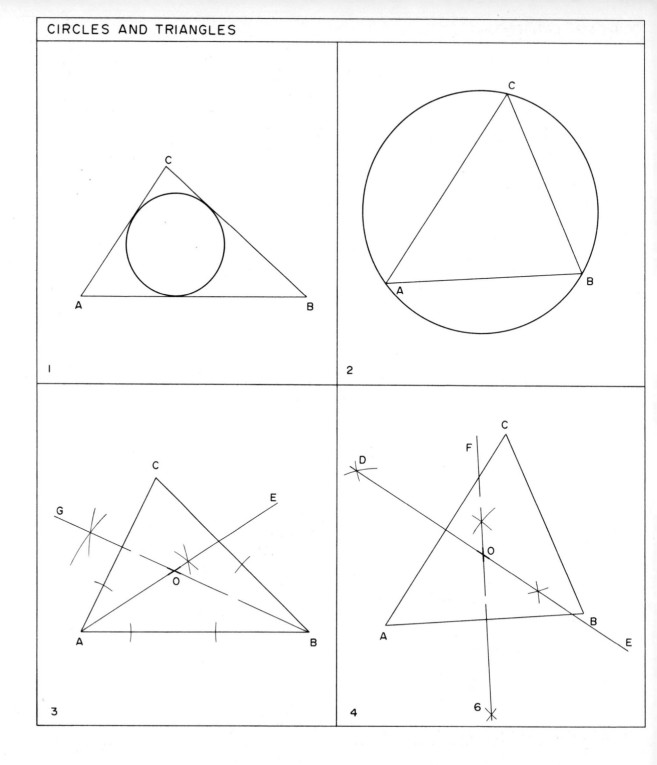

CIRCLES AND TRIANGLES

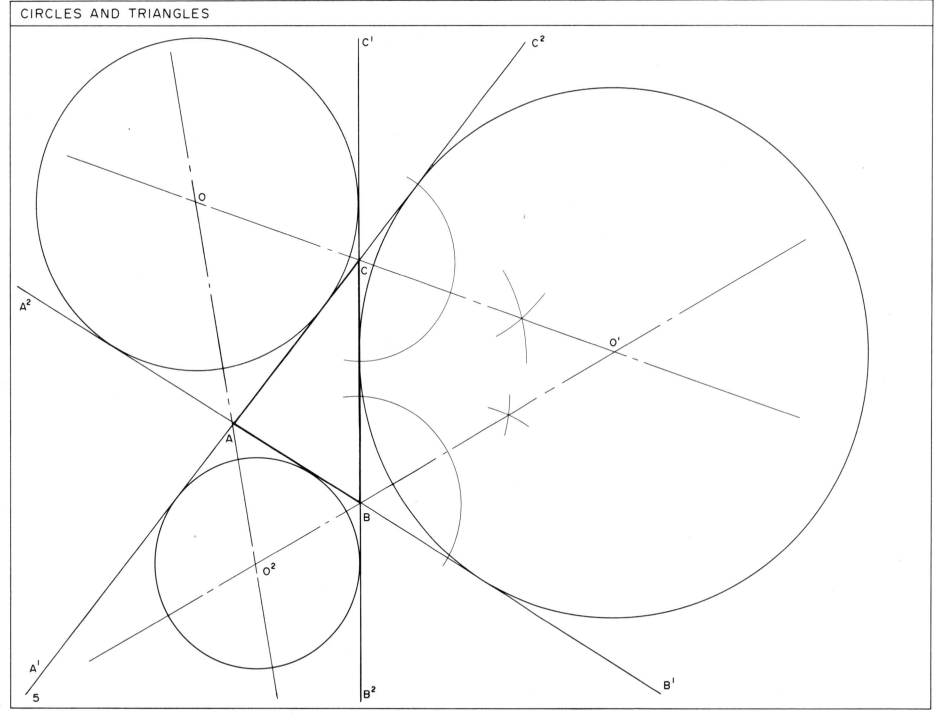

LOCI

A locus is the path traced out by a point moving in accordance with a definite law.

A mechanism consists of a set of links and rods which move together in ordered and regular manner.

The method of showing movement is to draw the mechanism in several different positions marking the point of the required locus each time.

To plot the locus of a point in a mechanism.
Draw this exercise following, stage by stage, the instructions and the drawings in the panels.

Panel No:

1. This shows a mechanism consisting of a crank, a connecting rod and a slider.

    The crank rotates counter clockwise and the slider is constrained to move along the line XX.

    We are going to draw the locus of point P, which is midway along the connecting rod.

2. Draw line XX. With your compasses set to 50 mm (the length of OA) draw a circle. Divide the circle into twelve equal parts and number them as shown.

3. With your compasses set at 130 mm (the length of AB) place their point at position No. 1 on the circle and strike an arc to intersect XX at 1^1.

    Join 1 and 1^1 with a straight line and mark point P^1 midway along it.

4. With your compasses still set at 130 mm place their point at position No. 2 on the circle and strike an arc to intersect XX at 2^1.

    Join 2 and 2^1 with a straight line and mark point P^2 midway along it.

5. With your compasses still set at 130 mm place their point at position No. 3 on the circle and strike an arc to intersect XX at 3^1.

    3 and 3^1 are already joined by line XX. Mark points P3 midway between 3 and 3^1.

6. Repeat the procedure for all twelve points on the circle then join the mid points (P^1 P^2 P^3 etc.) with a smooth curve.

    This curve is the locus (or the path) of P for one revolution of the crank OA.

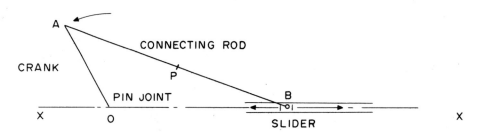

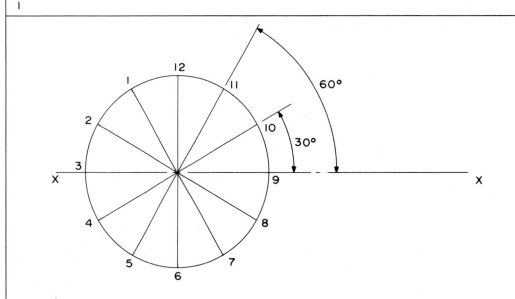

LOCI

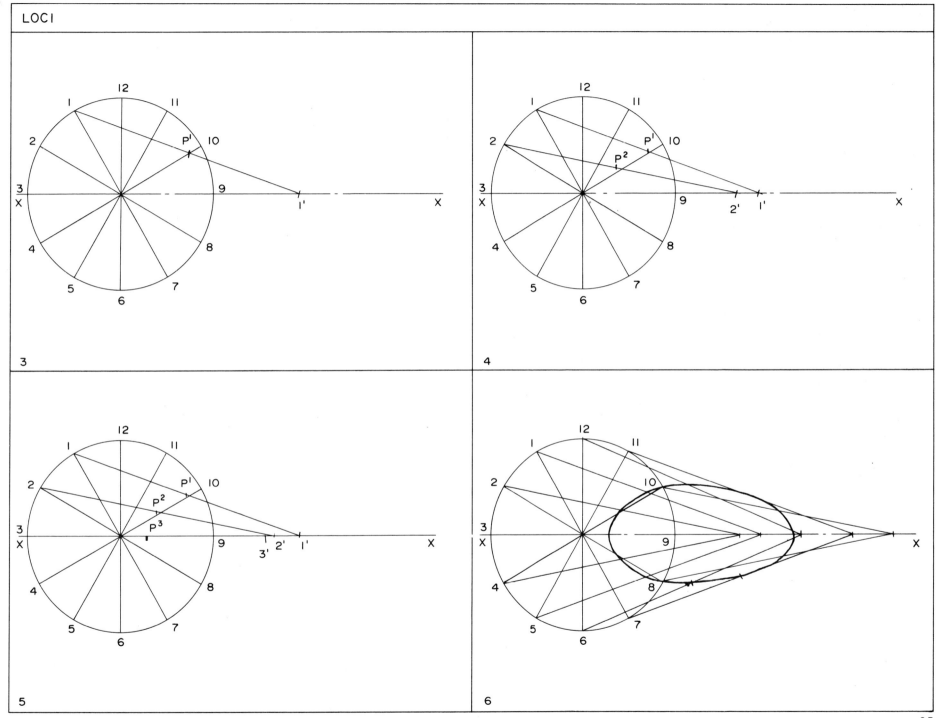

To plot the locus of a point in a mechanism.
Draw this exercise following, stage by stage, the instructions and the drawings in the panels.

Panel No:

1. In the mechanism shown AB is a crank which rotates about A.

   The rod BC is pin jointed at B and is free to slide through the guide D.

   The guide D remains fixed but oscillates to allow rod BC to slide freely.

   We are going to plot the locus of C for one revolution of the crank AB.

2. Draw the circle (radius 50 mm) which represents the path of the crank AB.

   Fix the position of the slider D to the dimensions given.

   The rod BC is 140 mm long.

3. Divide the circle into twelve equal parts and mark each one B1, B2, B3 etc. These indicate different positions of B.

4. From B1, draw a straight line passing through D and from B1 measure 140 mm to fix the position of C1.

5. From B2, draw a straight line passing through D and from B2 measure 140 mm to fix the position of C2.

6. Repeat the process for B3, B4 and B5.

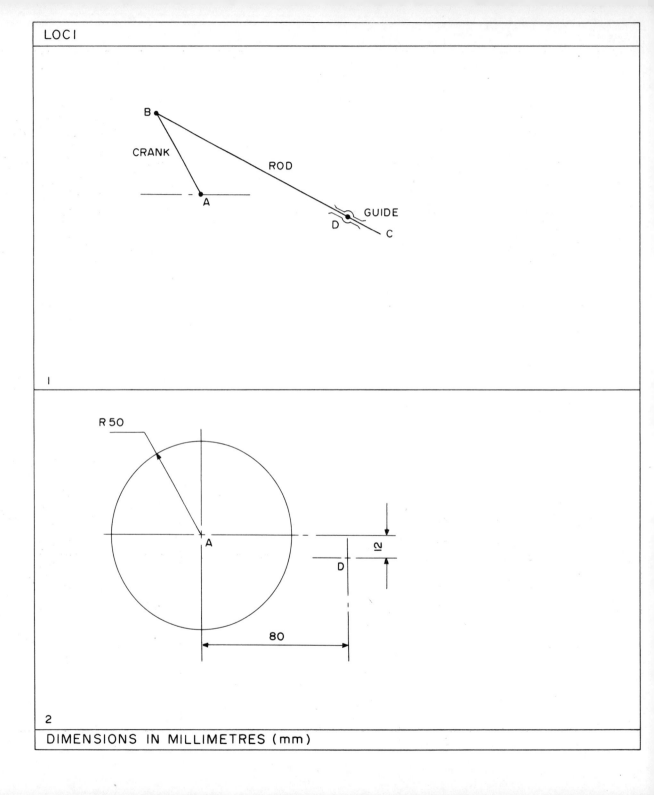

DIMENSIONS IN MILLIMETRES (mm)

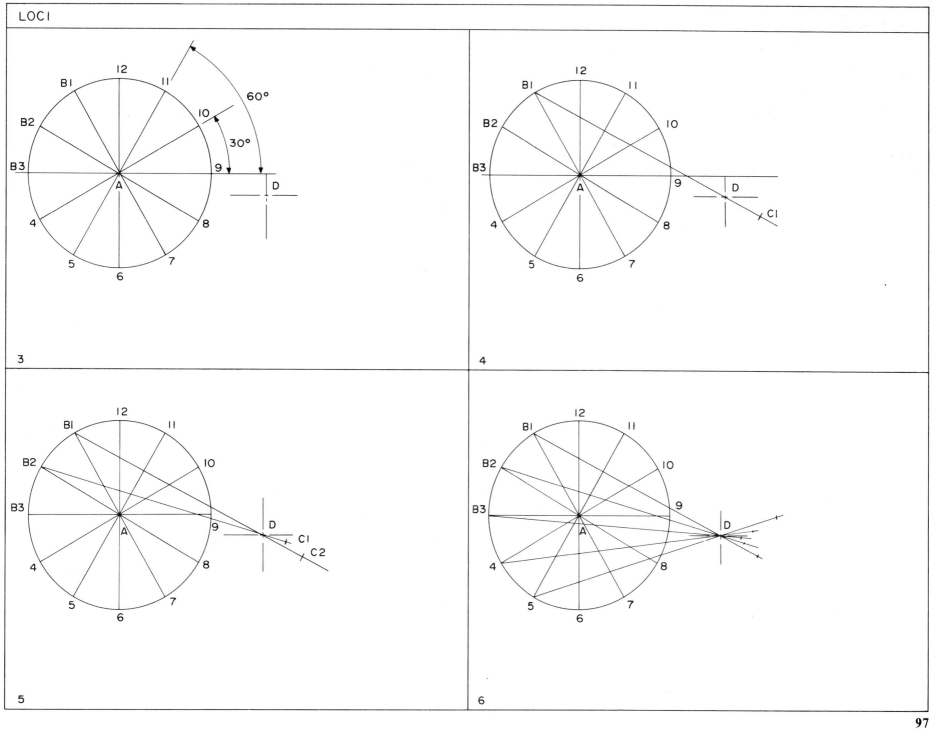

7-8 These panels show the remaining points plotted. Note the gap between C8 and C9.

9 This panel shows additional positions of the crank and connecting rod (8a and 9a) which have been inserted to give additional positions of C (C8a, C9a) to close the gap between C8 and C9.

10 Complete by joining the positions of C with a smooth curve.

Exercise:

11 Plot the locus of point C for one revolution of the given mechanism.

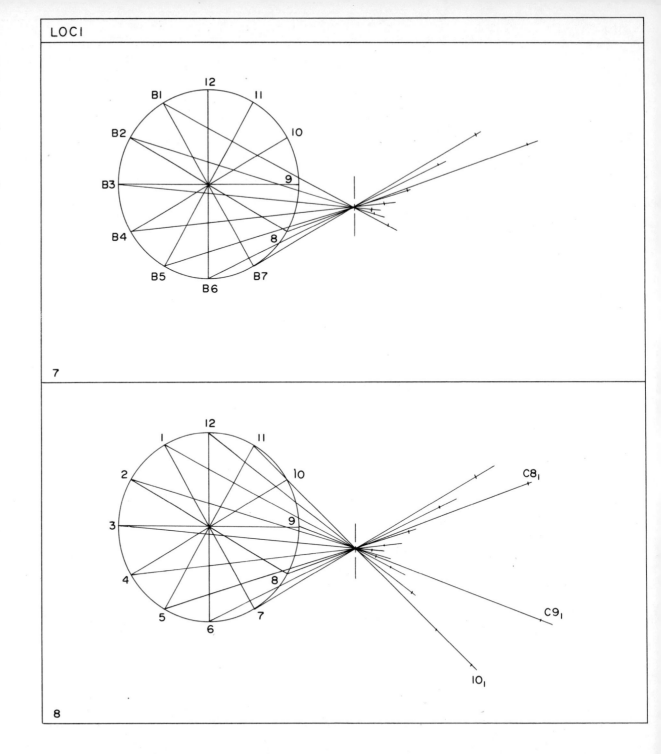

LOCI

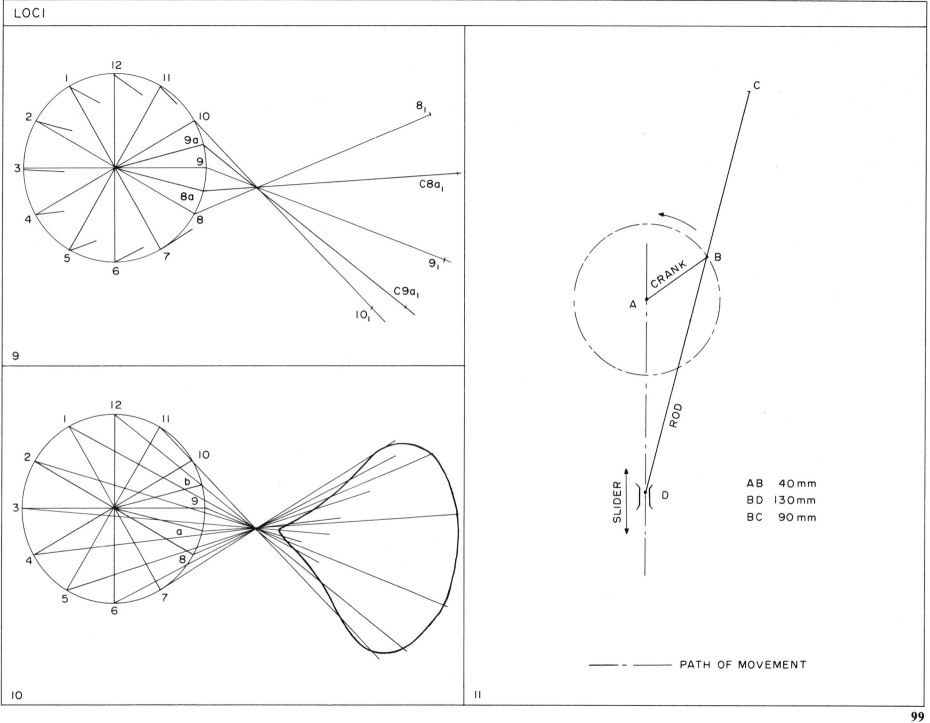

The involute is the path of a point on a straight line which rolls (without slipping) around a circle.

It can best be visualized by imagining a piece of string wound round a cylinder. If the string is unwound and kept taut the free end will trace an involute.

To draw the involute of a circle of 50 mm dia.
Draw this exercise following, stage by stage, the instructions and the drawings in the panels.

Panel No:

1. Draw a 50 mm dia. circle and divide it into twelve parts which should be numbered.

2. At point 1 draw a tangent and along it mark off 1/12 of the circumference of the circle. The length of the chord between any two points on the circle is a sufficiently close approximation for our purpose.

3. At point 2 draw a tangent and along it mark off 2/12 of the circumference.

4. At point 3 draw a tangent and along it mark off 3/12 of the circumference of the circle.

5. At point 4 draw a tangent and along it mark off 4/12 of the circumference of the circle.

6, 7 Shows you how to obtain all twelve points to draw the involute.
(Note: The true distance of point 12 from the circle is equal to its circumference).

8. Join all twelve points with a smooth curve.

Note the curve begins at point 12 on the circle.

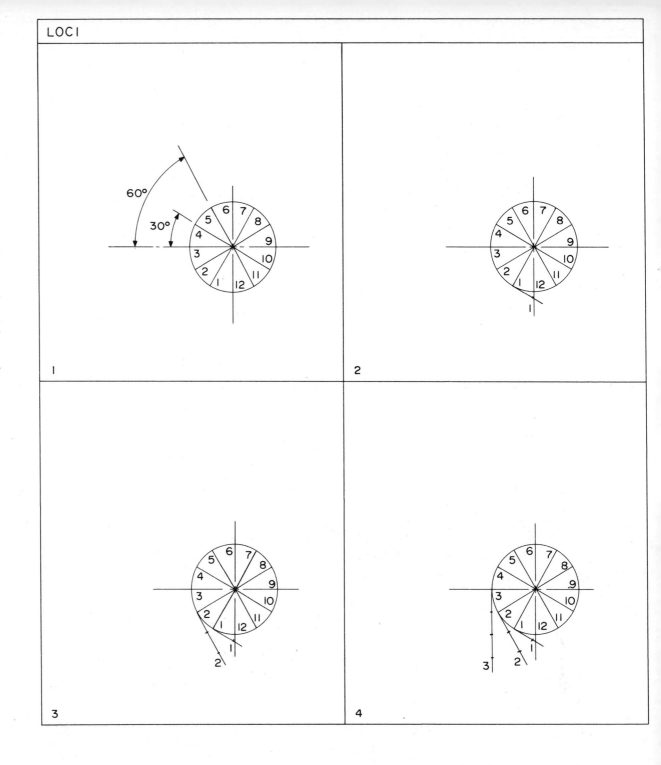

LOCI

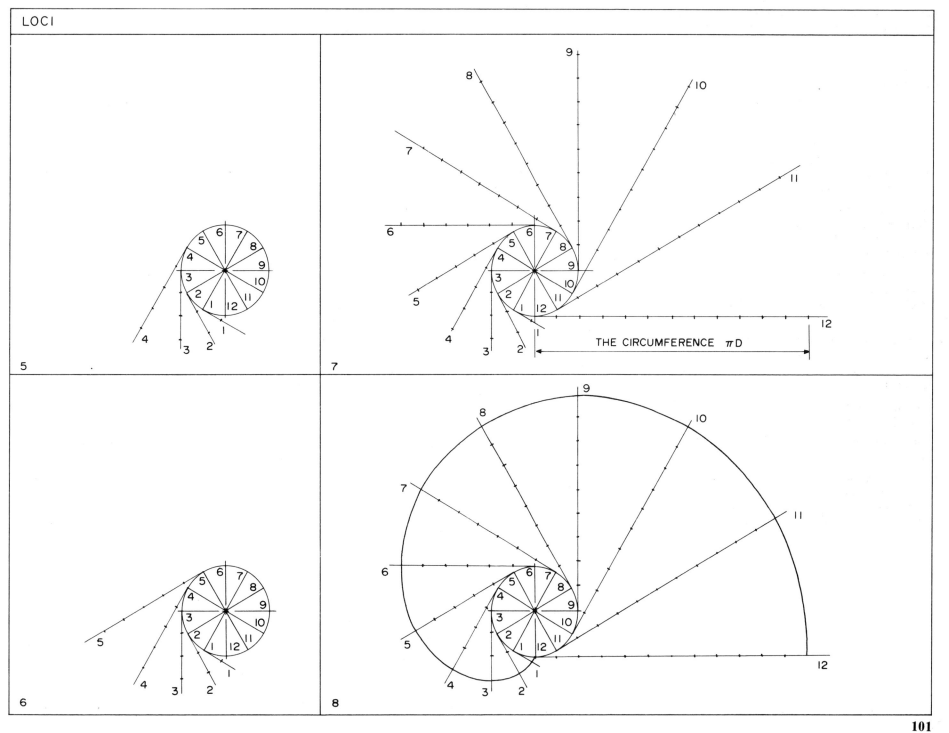

The cycloid is the locus of a point on the circumference of a circle which rolls, without slipping, on a straight line.

To draw a cycloid.
Draw this exercise following, stage by stage, the instructions and the drawings in the panels.

Panel No:

1. Draw a circle of 40 mm radius, draw a base line and divide the circumference of the circle into twelve equal parts. Number these parts. Part O is the start of the cycloid.

2. Mark one twelfth of the circumference along the base line and with centre O^1 draw the circle in position No. 1. Project point 1 as shown to give the next point on the cycloid.

3. Mark a further one twelfth of the circumference along the base line and with centre O^2 draw the circle in position No. 2. Project point 2 as shown to give a further point on the cycloid.

4-6 The circle is drawn in further positions, moving along the base line one twelfth of the circumference for each position, and points 3, 4 and 5 obtained as shown.

7. This shows the uppermost point on the cycloid and represents half the locus.

   The remaining six points are obtained in the same manner as the first seven.

8. Draw a freehand curve through the points obtained to obtain the cycloid as shown.
   (Note: For one revolution of the rolling circle the true distance travelled is its circumference).

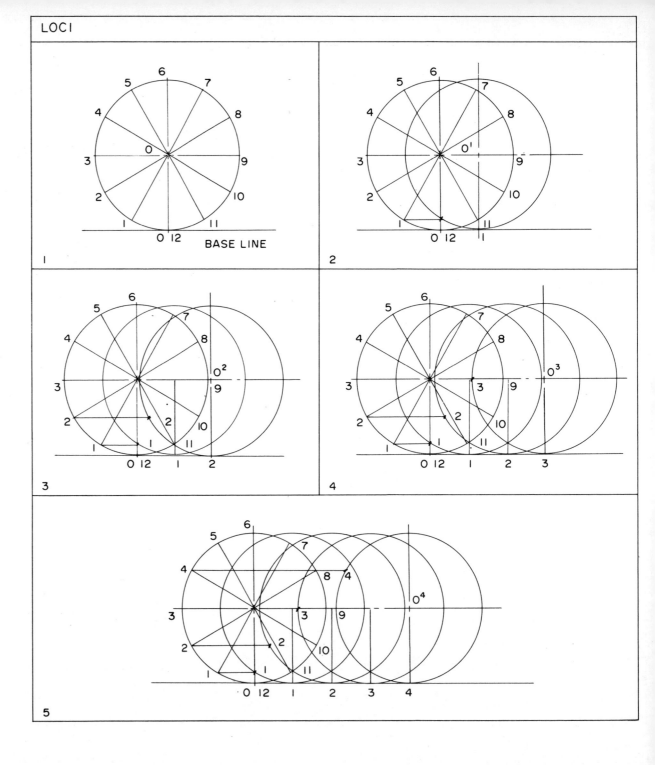

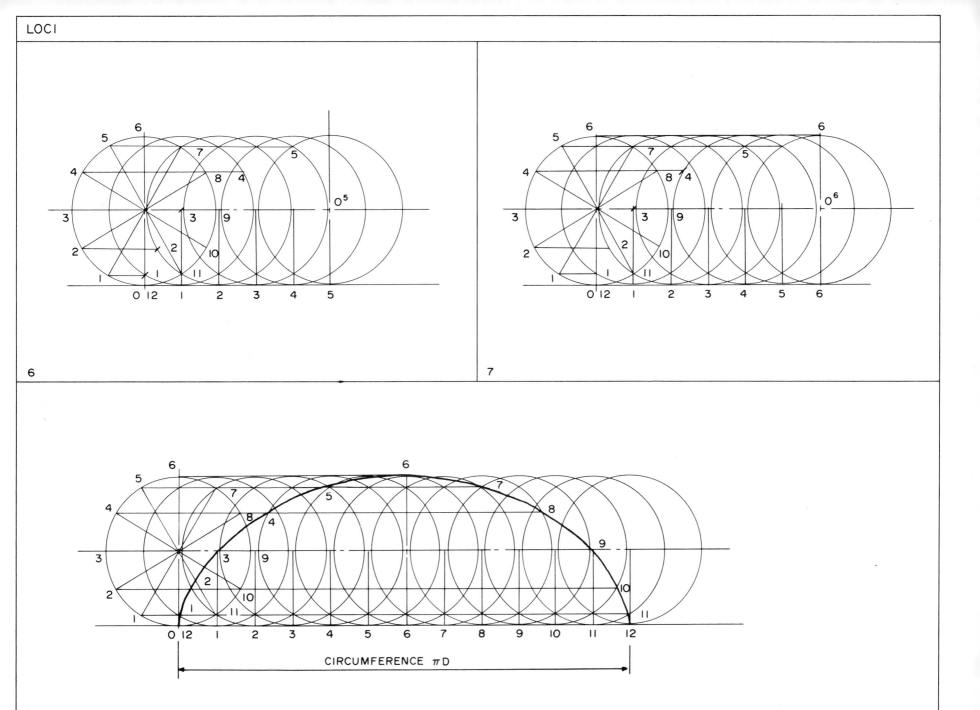

The locus of a point which is always equidistant from a given line and a given fixed point is a parabola.

To draw the locus of a point equidistant from a given fixed point and a straight line.
Draw this exercise following, stage by stage, the instructions and the drawings in the panels.

Panel No:
1. This shows a point P on the required curve. F is the fixed point.
$$PF = PD \text{ or } \frac{PF}{PD} = 1$$

2. Draw the line AB and point F to the given dimensions.

3. Bisect DF to give point V.

4. From V mark a series of points 1, 2, 3, 4 etc., approximately 6 mm apart.

5. Through point 1 draw a straight line parallel to line AB.

6. With the point of your compasses on F and set to radius of D1 draw arcs to intersect the line through point 1 as shown giving two points on the curve.

7. Through point 2 draw a straight line parallel to line AB.

8. With the point of your compasses on F and set to a radius of D2 draw arcs to intersect the line through point 2 as shown giving two more points on the curve.

9. Through point 3 draw a straight line parallel to line AB.

10. With the point of your compasses on F and set to a radius of D3 draw arcs to intersect the line through point 3 as shown giving two more points on the curve.

    Join the points obtained by a freehand curve.

    To obtain more of the curve the process can be repeated through further points 6, 7, 8, 9, etc.

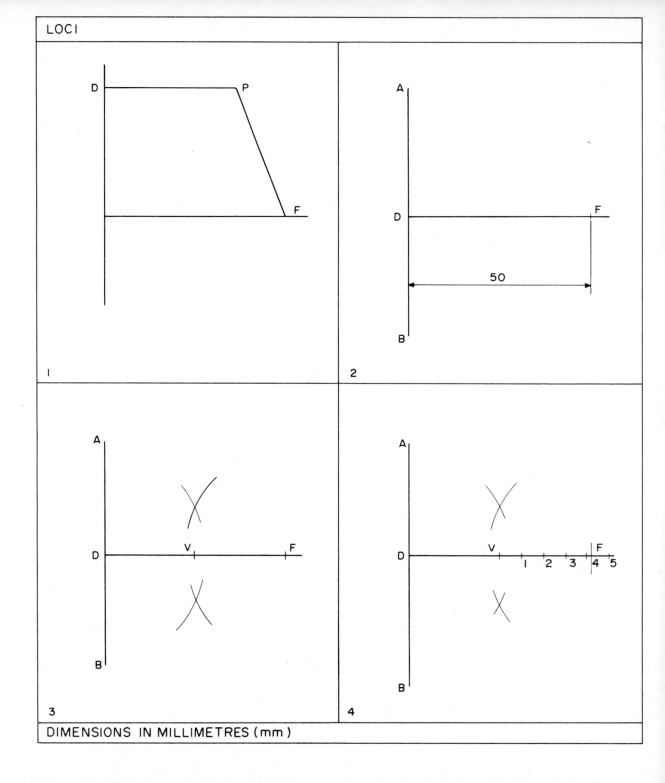

LOCI

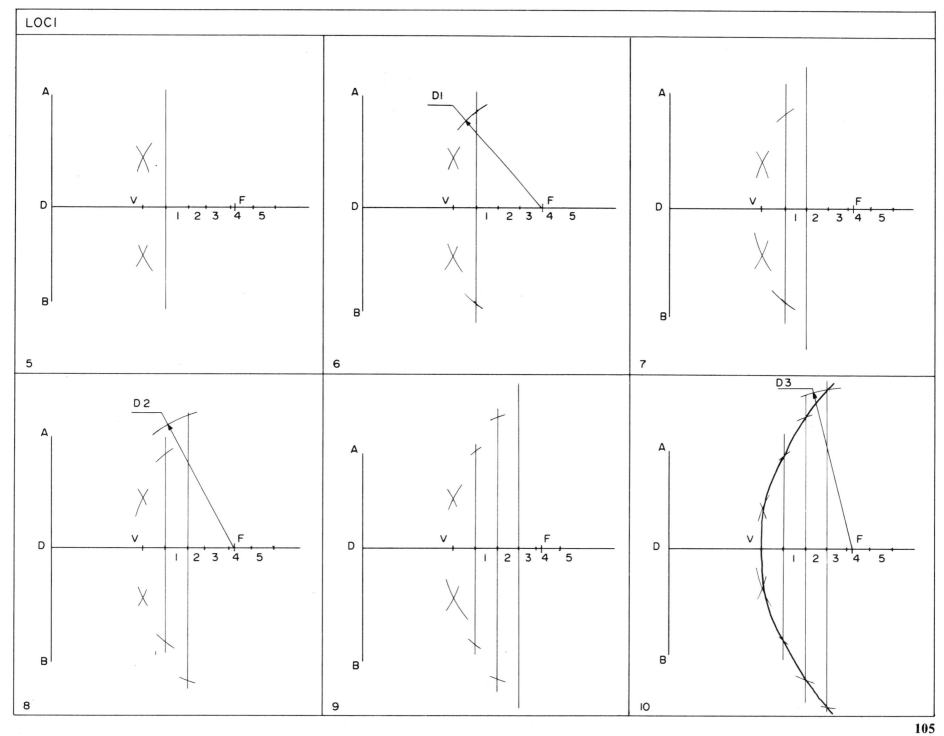

The helix

This is an important locus being the basic form of the screw thread. It can be regarded as a line of uniform gradient on a cylinder.

Panel No:

1. This shows a cylinder with its side view and plan from above. The right angled triangle ABC has its side CB of uniform gradient.

2. The triangle ABC is now wrapped round the cylinder and its position shown by the letters $A^1 B^1 C^1$. The straight line BC is now formed into a curve called a helix. After the locus has completely encircled the cylinder the distance it moves along the axis of the cylinder is known as the pitch.

To draw one turn of a helix to given dimensions.

Draw this exercise following, stage by stage, the instructions and the drawing in the panels.

Panel No:

3. Draw the two views of the cylinder to the given dimensions.

4. Divide the plan into eight equal parts.

5. Divide the length of the cylinder into eight equal parts, in this case to be the pitch of the helix.

6. Project from point 1 in the plan to 1 in the side view.

7. Continue with point 2 in the plan to 2 in the side view.

8. Proceed as shown with point 3.

9. This shows the completed helix with part of the curve behind the cylinder drawn in hidden detail.

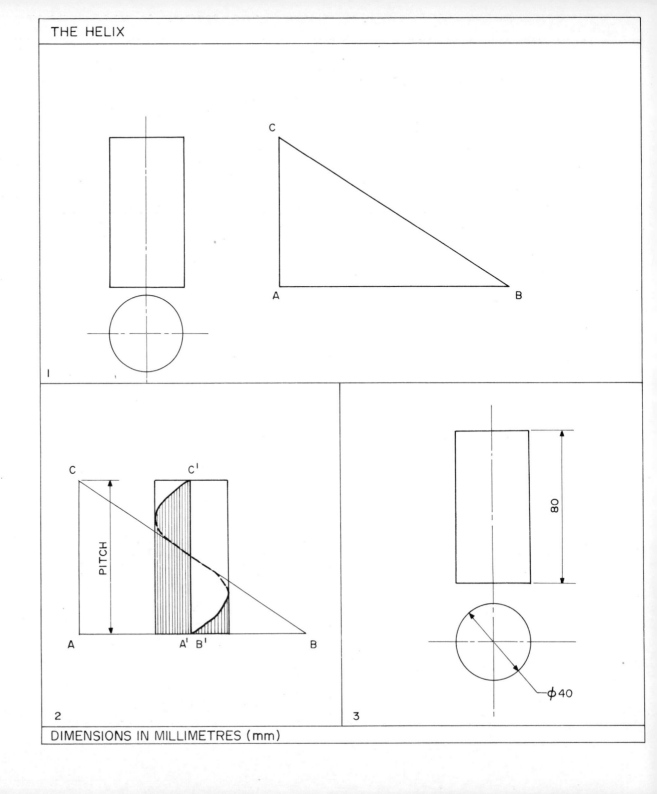

DIMENSIONS IN MILLIMETRES (mm)

THE HELIX

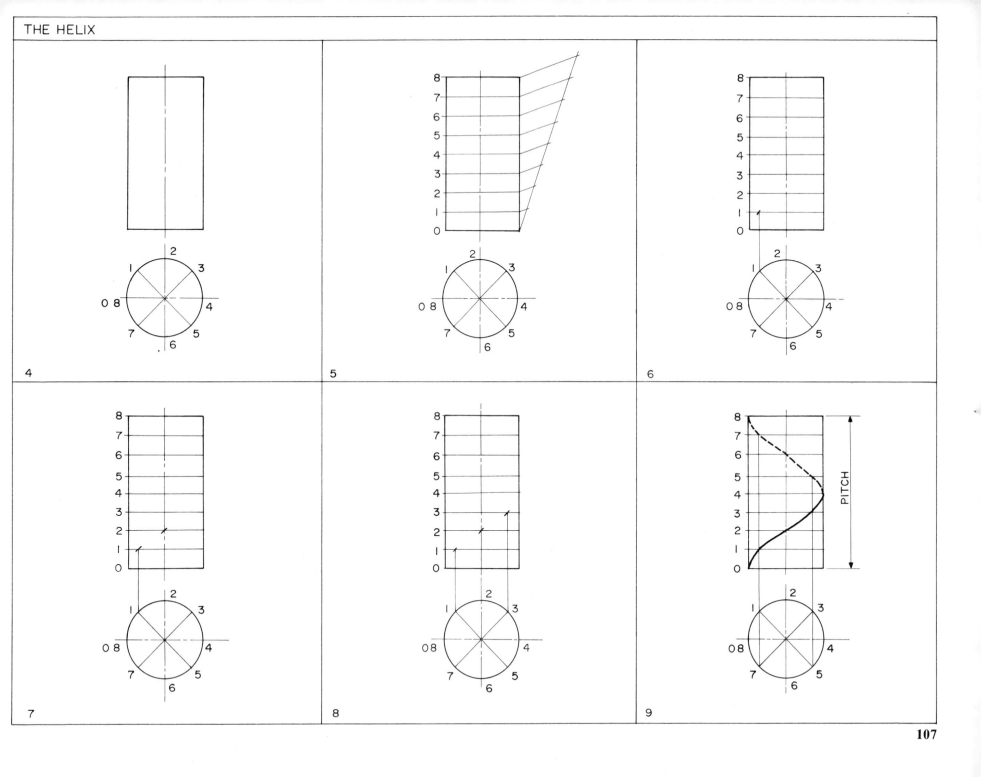

The ellipse

There are many ways in which an ellipse can be obtained. It can be regarded as the locus of a point that moves in such a manner that the sum of its distances from two given fixed points remains constant. The following construction is based upon this property of the ellipse.

To draw an ellipse with a major axis of 120 mm and a minor axis of 70 mm.

Draw this exercise following, stage by stage, the instructions and the drawings in the panels.

Panel No:

1. Draw the two axes AB = 120 mm and CD = 70 mm.

2. Set your compasses to half the length of the major axis (OB or OA) and with the point at D, strike arcs to intersect the major axis giving F_1 and F_2. These are the focal points of the ellipse.

3. Mark a series of numbered points between O and F_1 making sure that the last one is close to F_1.

4. Set your compasses to a radius of A→1 and with the point on F_1 draw an arc as shown.

5. Set your compasses to a radius of 1→B and with the point of F_2 draw an arc to intersect the one previously drawn from F_1. This fixes one point on the ellipse.

6. Using the two radii A→1 and 1→B arcs drawn from F_1 and F_2 will fix four points in the ellipse.

7. Set your compasses to a radius of A→2 and with their point on F_1 draw arcs GH, G^1H^1. Without altering your compasses use F_2 as the centre to draw arcs G^2H^2 and G^3H^3.

8. Set your compasses to a radius of 2→B and with their point on F_2 draw arcs to intersect arcs GH and G^1H^1 then with their point on F_1 draw arcs to intersect arcs G^2H^2 and G^3H^3 giving four more points on the ellipse.

9. Repeat the process for remaining points. If necessary additional points can be inserted to ensure a better curve.

10. A freehand curve drawn through the points obtained gives the required ellipse.

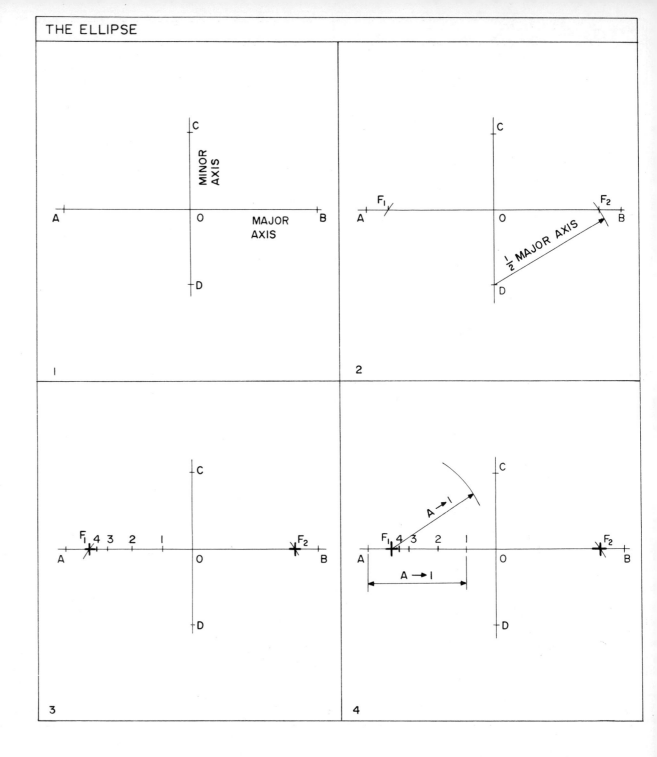

THE ELLIPSE

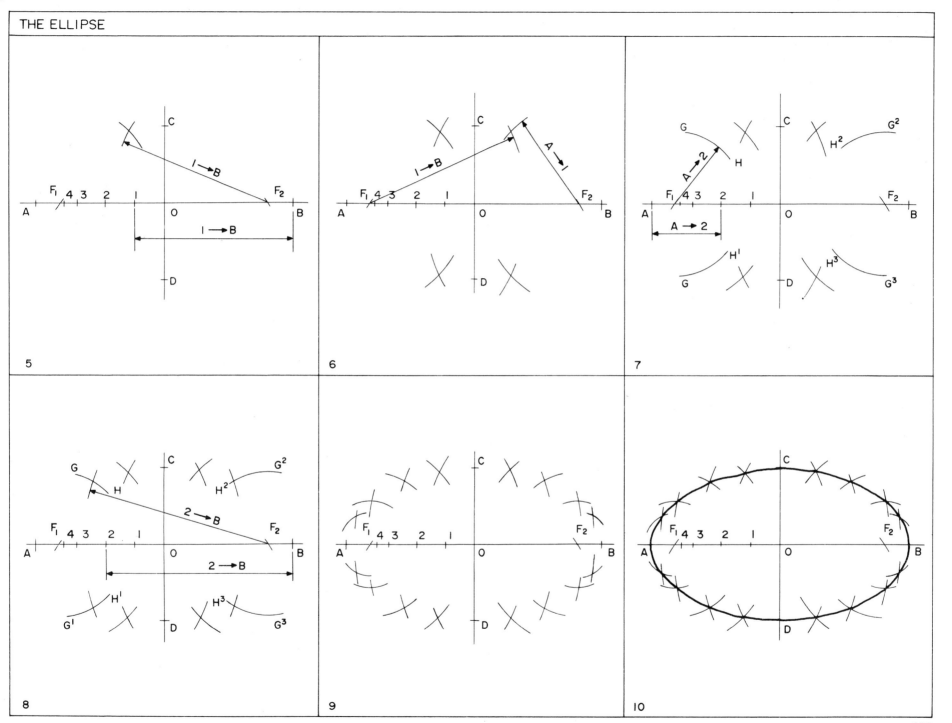

To draw an ellipse major axis 100 mm, minor axis 60 mm using the concentric circle construction.

Draw this exercise following, stage by stage, the instructions and the drawings in the panels.

Panel No:

1. Draw two lines at right angles, crossing at O. These will form the axes of the ellipse.

2. Using O as centre draw a circle of 100 mm dia. AB is the major axis of the ellipse.

3. Using O as centre draw a circle of 60 mm dia. CD is the minor axis of the ellipse.

4. Through O draw line EF to cut both circles.

5. Through E^1 and F^1 draw horizontal lines.

6. Through E and F draw vertical lines to intersect the horizontal lines from E^1 and F^1. This fixes two points on the ellipse.

7. Through O draw line GH to cut both circles and repeat the process to give two more points on the ellipse.

8. Draw further lines to obtain sufficient points to draw the ellipse.

9. A freehand curve drawn through the points obtained results in an ellipse of the required dimensions.

10. This panel suggests the angles at which the lines EF, GH, etc., can be drawn.

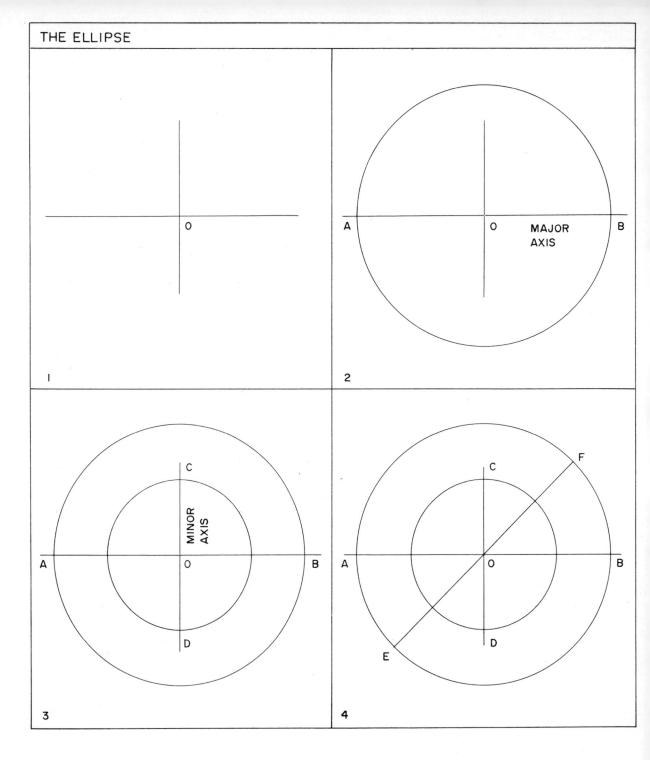

THE ELLIPSE

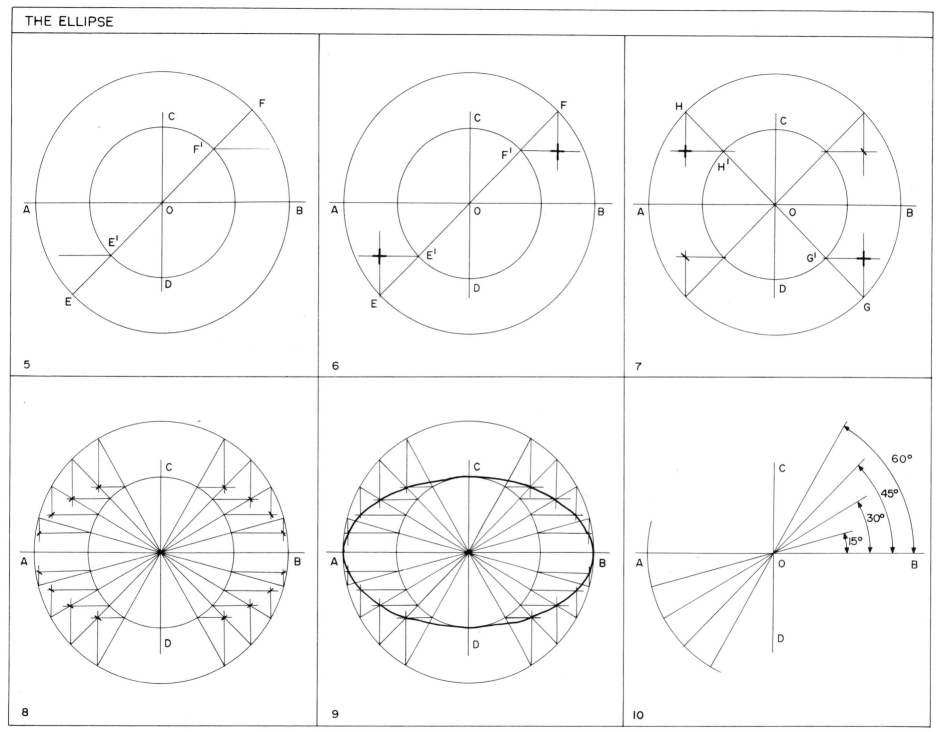

To draw the locus of a point which moves in such a manner that it is always equidistant from a given fixed point and the circumference of a given circle.

Draw this exercise following, stage by stage, the instructions and the drawings in the panels.

Panel No:

1. Draw the given circle, centre B and the fixed point A to the given dimensions.

2. Bisect the distance between point A and the circumference of the circle. Point P is equidistant from point A and the circumference of the circle and is one of the points in the locus.

3. With your compasses set to any convenient distance set out points 1 and 1^1 either side of point P.

4. With the point of your compasses on A and set to a radius of AP→1 draw arcs EF and E^1F^1.

5. With the point of your compasses on the centre of the circle B and set to a radius of BP→1^1 draw arcs GH and G^1H^1 to intersect arcs EF and E^1F^1. The intersections of the arcs fix two points P^1 and P^2 in the locus.

6. With your compasses set to any suitable measurement and with the point on P mark off points 2 and 2^1 either side of P.

7. With the point of your compasses on point A and set to a radius of AP→2 draw arcs JK and J^1K^1.

8. With the point of your compasses on the centre of the circle B and set to a radius of BP→2^1 draw arcs LM and L^1M^1 to intersect arcs JK and J^1K^1 fixing points P^3 and P^4 in the locus.

9. With your compasses set to any suitable measurement and with the point on P, mark off points 3 and 3^1 on either side of P.

    With the point of your compasses on A and set to a radius of AP→3 draw arcs NO and N^1O^1.

10. With the point of your compasses on the centre of the circle B and set to a radius of BP→3^1 draw arcs PQ and P^1Q^1 to intersect arcs NO and N^1O^1 fixing points 5 and 6 in the locus.

    Draw a freehand curve through points 5, 3, 1, P, 2, 4, 6 to give the required locus.

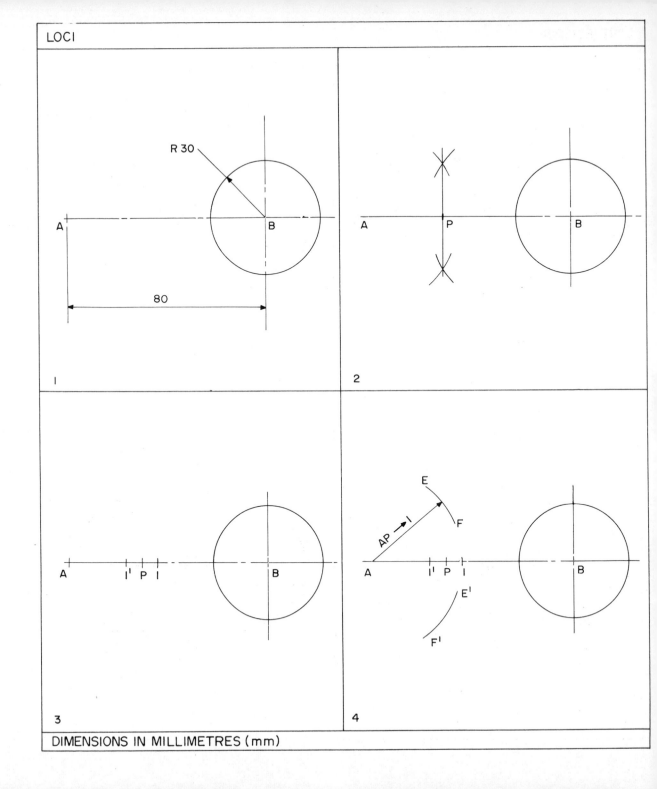

LOCI

DIMENSIONS IN MILLIMETRES (mm)

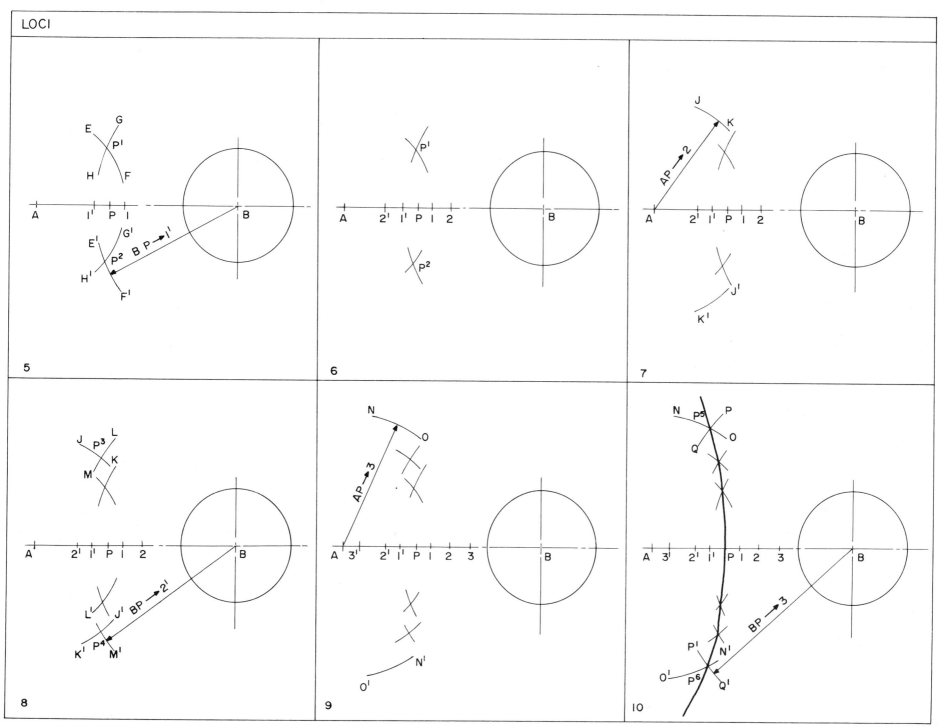

ORTHOGRAPHIC PROJECTION

Orthographic projection is the geometric basis of machine drawing. A solid which has three dimensions is represented on paper which has two.

Panel No:

1. This shows three planes (flat surfaces) meeting like the corner of a box. The planes are at right angles to each other and are named: Vertical Plane (V.P.) Horizontal Plane (H.P.) and Side Vertical Plane (S.V.P.).

   Three views of a shaped block placed in space enclosed by the planes can be projected onto them as shown.

2. The block has been omitted to show the relationship of the views to each other.

3. This shows the usual layout for orthographic projection and the names of the views. The XY lines are the hinge lines of the unfolded planes. The change of direction line enables projection from plan to end elevation (or vice versa).

4. Compasses can also be used to project between plan and end elevation.

Draw in orthographic projection the object shown in Panel No. 5 following stage by stage the instructions and the drawings in the panels.

Panel No:

6. Draw the XY and X_1Y_1 lines and change of direction line CD as shown and the centre lines (chain lines) to the given dimensions.

7. Complete the compass work as shown.

8. Draw the base of the front elevation and project into the H.P. and the S.V.P.

9. Set out the plan and project into the S.V.P.

10. Visible edges are lined in and hidden edges shown by a broken line. The symbol at the top of the drawing shows the projection used.

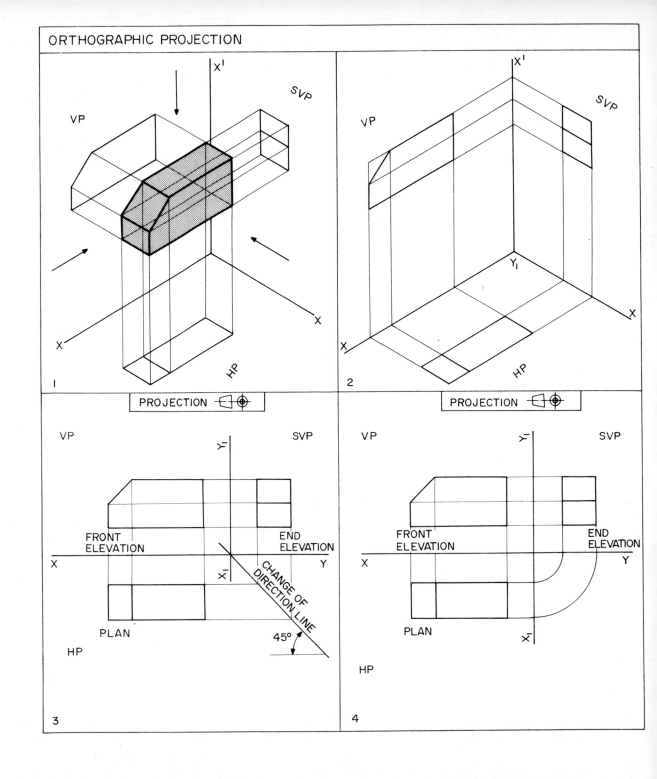

114

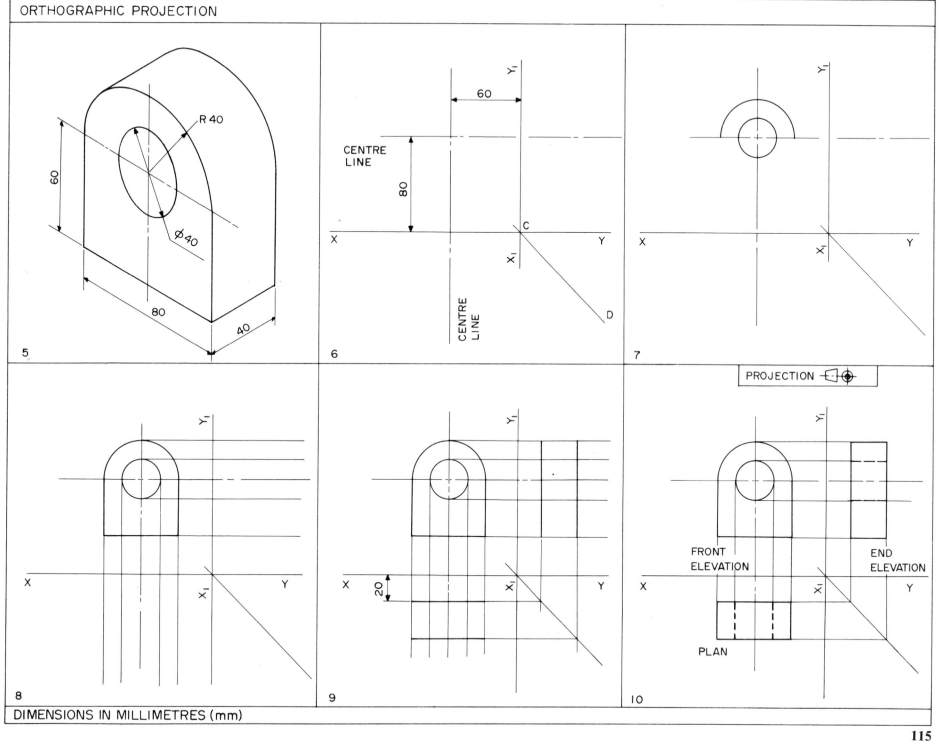

Any solid, regardless of its position, can be drawn in orthographic projection. In the following examples a 30 mm cube in various positions is to be shown as a plan, front elevation and an end elevation in each case.

Panel No:
1. Draw the plan of the cube and XY lines as shown including the change of direction line.
2. Project the corners of the plan into the elevation.
3. The corners of the plan are projected to the change of direction line.
4. Project from the change of direction line into the end elevation.
5. Mark the height of the cube 'h' in the elevation; project and line in as shown.
6. Line in the visible vertical edges in the elevations and show the hidden edges by a broken or pecked line.

Exercises:
7-10 Using the above procedure draw a plan, front elevation and end elevation for each example. Include hidden detail.

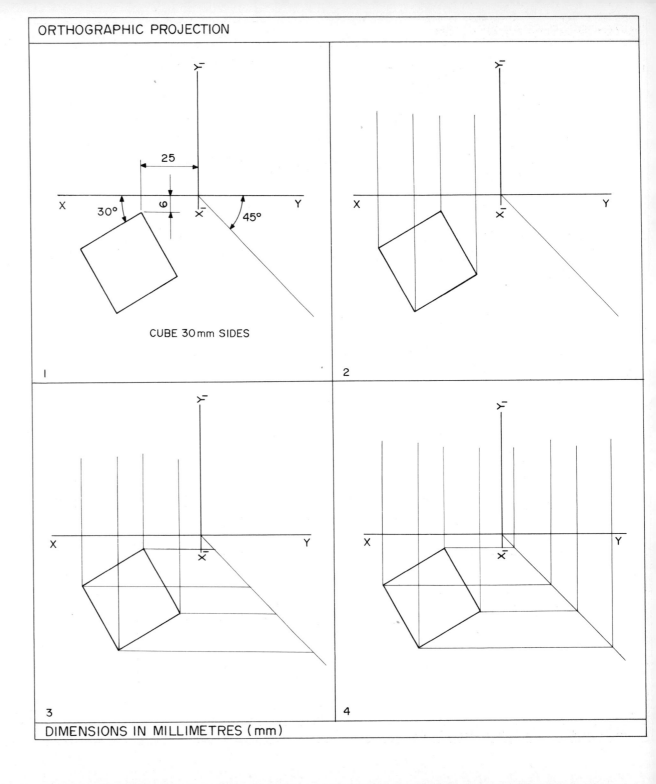

116

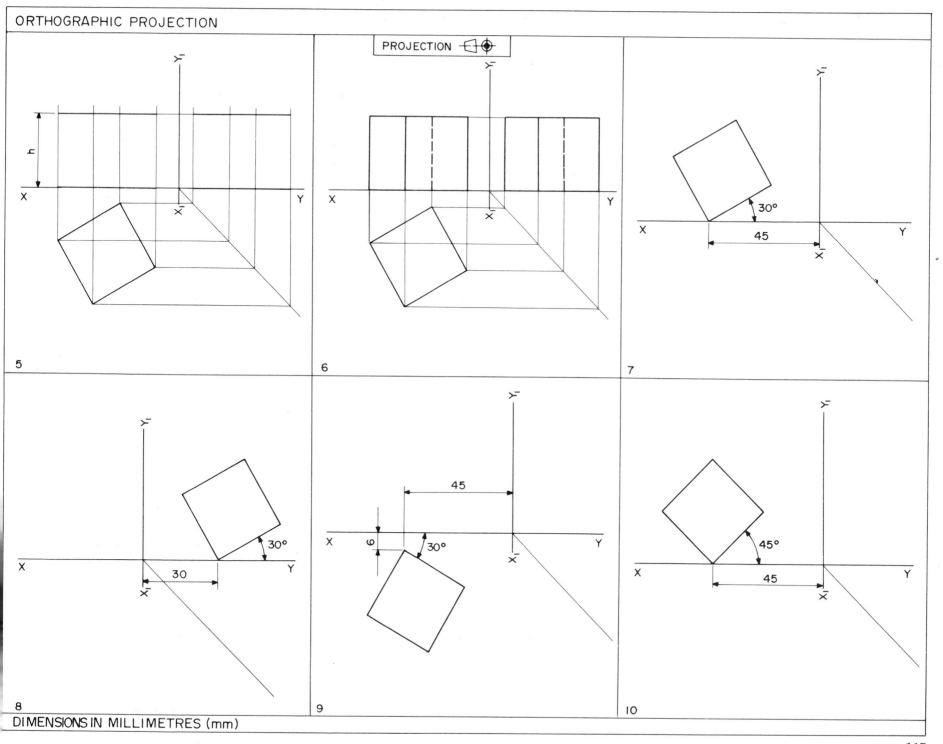

For a drawing to be read successfully it is necessary to set out views from the most informative direction and include a sufficient number of views to avoid ambiguity.

Panel No:

1-4 In these four examples the elevations are exactly the same, the shape of the object only becomes clear when the plan is given.

(Note: It is not always necessary to provide three views, in these examples the front elevation and plan would suffice).

To project a view of a solid from given views.
Draw this exercise following, stage by stage, the instructions and the drawings in the panels.

Panel No:

5 This panel shows two elevations of a solid.

6 Make a freehand sketch approximately full size conforming as far as possible to the requirements of the given elevations.

7 Line in the shape of a realistic solid which would give the required elevations. There may be many alternatives.

8 Draw the plan and elevations of the sketched solid to the measurements given in panel 5.

9,10 Using the same method now draw possible plans for these two examples.

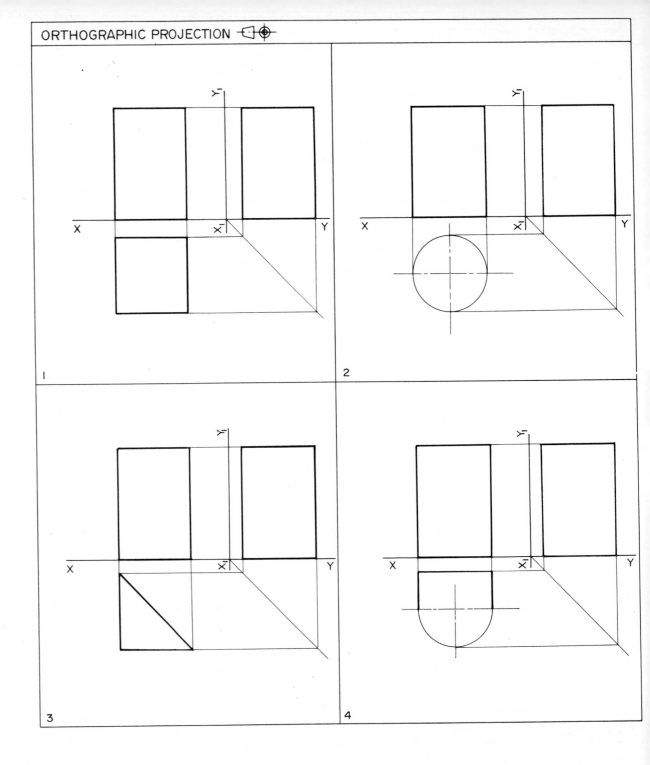

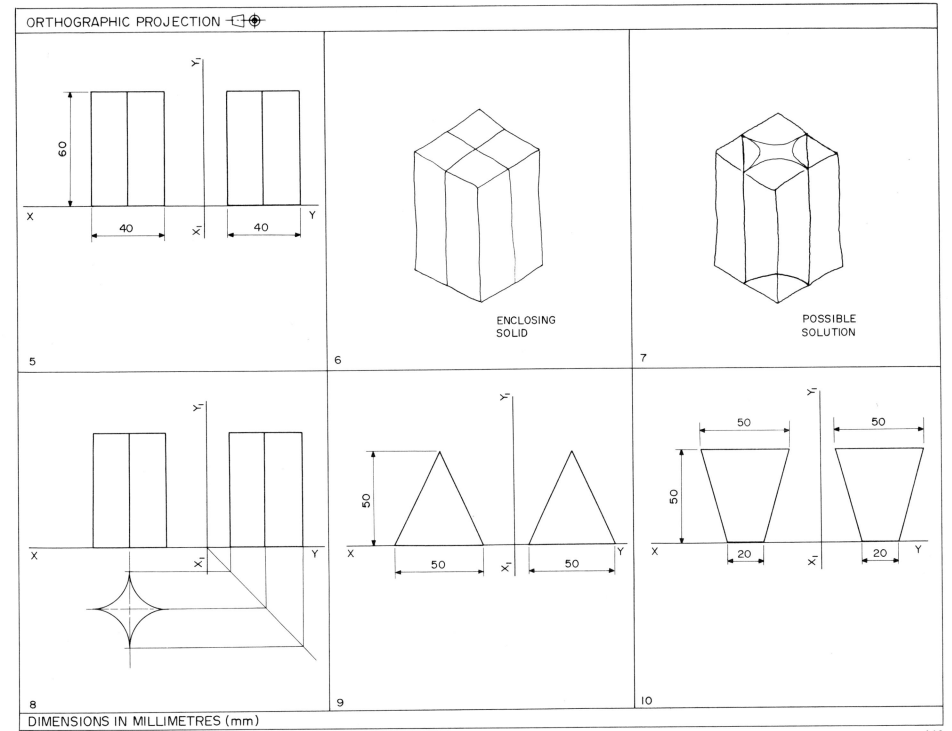

Third angle.
The examples we have drawn so far have been in first angle projection. Third angle is also widely used and the views set out as follows.

Panel No:
1. This illustrates the two planes of projection intersecting at right angles. The first angle is the one we have used so far and now we are to consider the third angle.

2. An object is shown enclosed by the planes of projection and can best be imagined as a transparent box through which the object is viewed in the direction of the arrows.

3. In this view the object is omitted to show the views in projection.

4. The planes are now shown unfolded to illustrate the normal layout for 3rd Angle projection.
   (Note: The simple way to remember this layout is that the view from above is drawn above and the view from the left is drawn on the left).

To draw the object shown in Panel No. 5 in 3rd Angle Projection.
Draw this exercise following, stage by stage, the instructions and the drawings in the panels.

Panel No:
6-10 These panels show the stages of your drawing leading to completion in Panel No. 10.

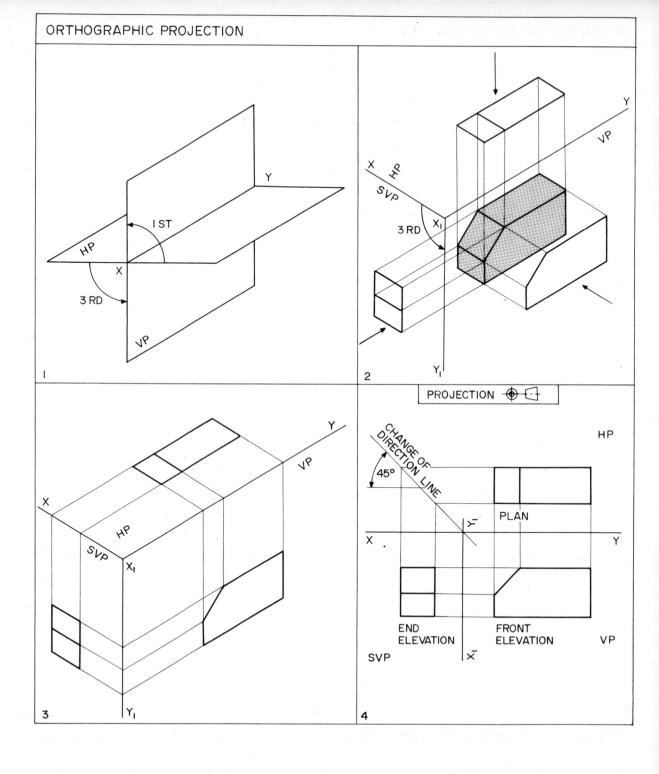

120

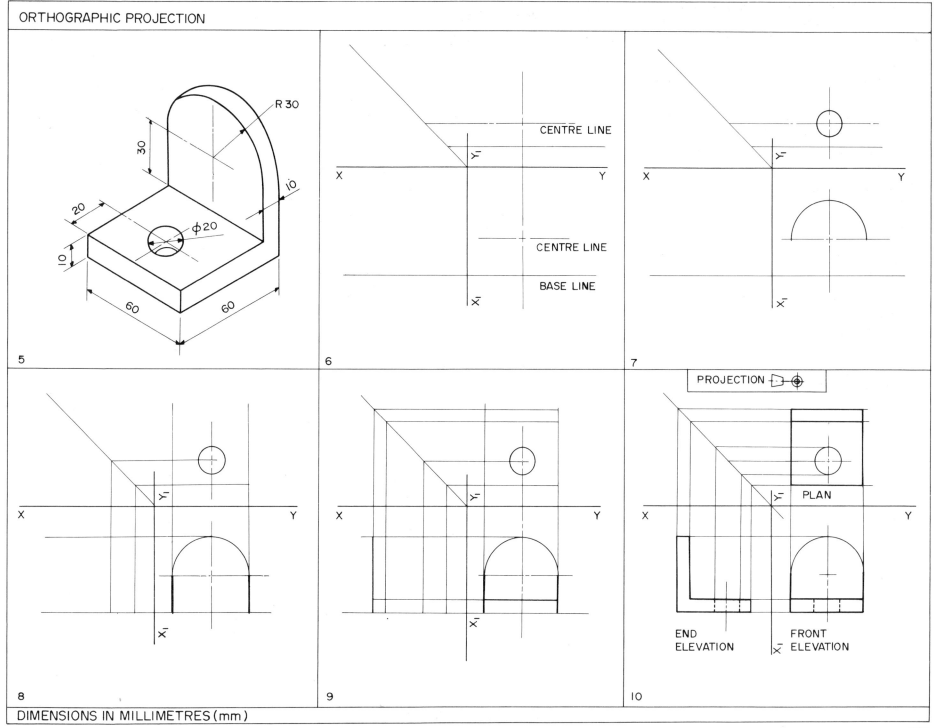

To draw an inclined solid in 3rd Angle Projection.
Draw this exercise following, stage by stage, the instructions and the drawings in the panels.

Panel No:

1. This shows the plan and elevation of a simple solid.

2. Using the dimensions of the solid in Panel No. 1 draw its new plan in the position shown.

3. Project all the corners into the front elevation. This method is suitable for any simple solid.

4. Draw the base line as shown for the two elevations.

5. Project the widths from the plan onto the change of direction lines.

6. Project from the change of direction line into the end elevation.

7. Mark the overall height of the solid and the step as shown.

8. Line in the base of both elevations.

9. Line in the vertical edges including hidden detail.

10. Complete the lining in of the horizontal edges in the elevations.

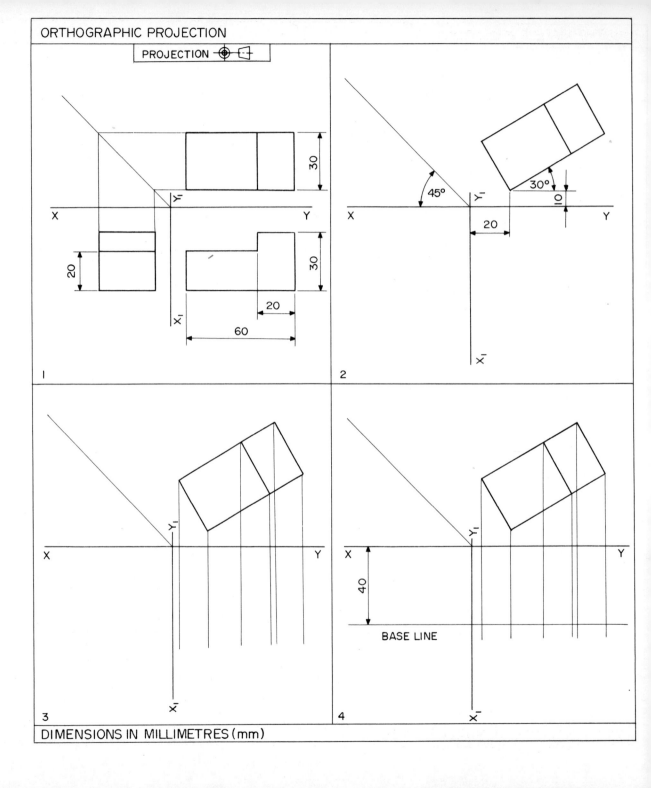

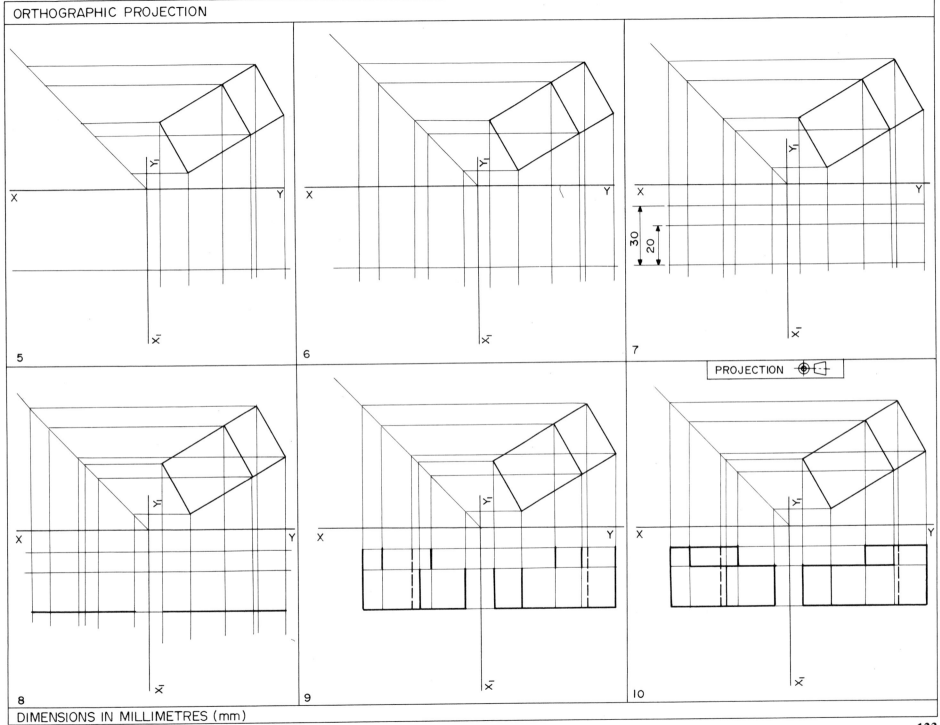

Sections.

Hidden detail on a drawing is frequently the source of confusion, the end elevation in Panel No. 6 is an example of this. A section or cut through a solid reveals the precise shape in that plane.

Panel No:

1. A block with a square hole is cut or sectioned on the plane indicated by the shading.

2. The portion to the left of the section has been removed to show the internal shape.

3. A sectioned drawing is started in the usual manner taking care that lining in is not affected by the section.

4. The chain line lettered AA in the plan shows the path or trace of the section. Any material cut by this imaginary line is shown by cross hatching in the elevation. These are lines at 45° drawn with the aid of a set square.

To draw in orthographic projection the following views of the object shown in Panel No. 5.
a) a plan,

b) a front elevation,

c) and end elevation sectioned on AA.

7-9 These panels show the procedure of drawing. The cross hatching is the final stage.

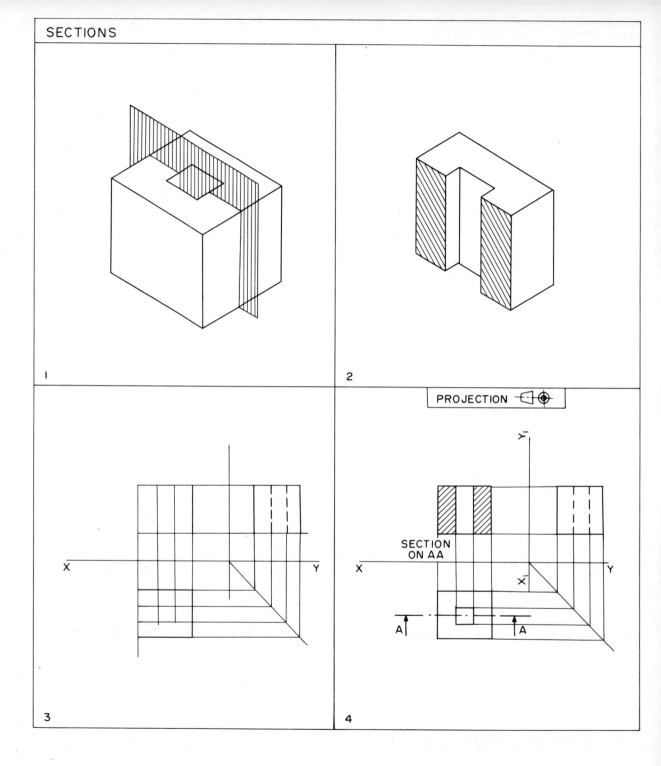

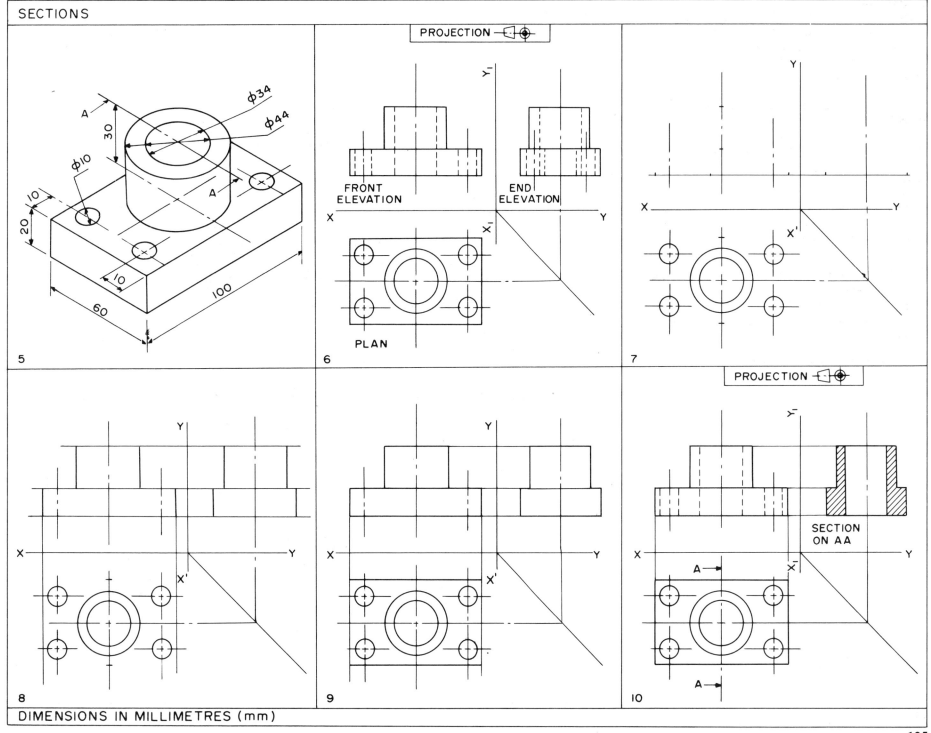

Auxiliary views

When part of an object is angled such as face A in the illustration in Panel No. 1, the true shape cannot be seen in the elevation or end elevation. An extra or auxiliary elevation becomes necessary with the plane of projection parallel to the face A.

To draw an auxiliary elevation of the block shown in Panel No. 1.

Draw this exercise following, stage by stage, the instructions and the drawings in the panels.

Panel No:

2. Draw the plan and elevation of the solid to the dimensions given, adding an extra XY parallel to the angled face in the plan. The change of direction line CD bisects the angle between the two XY lines and in this particular case is at 60°.

3. Project the corners of the plan of the solid at right angles to X_1Y_1.

4. Transfer the height 'h' of the solid into the auxiliary view by projecting parallel to the XY lines.

5. The auxiliary elevation is lined in and the true shape of face A is now part of the auxiliary view.

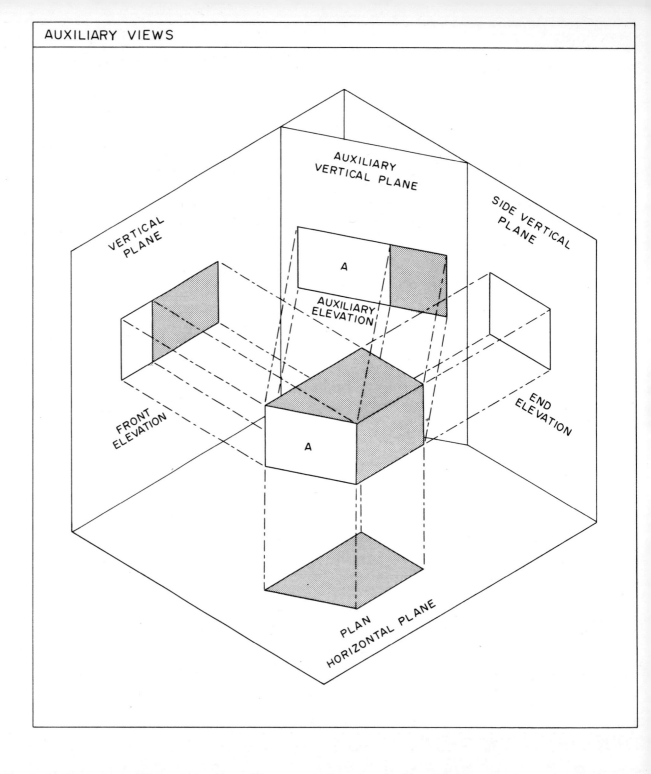

AUXILIARY VIEWS

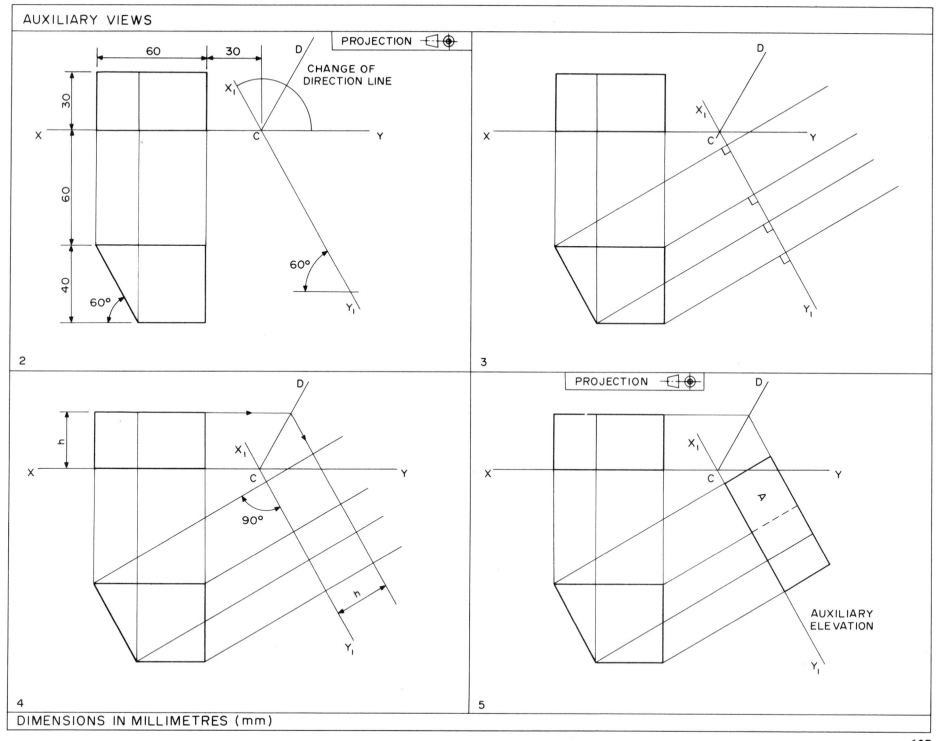

DIMENSIONS IN MILLIMETRES (mm)

In the illustration in Panel No. 1, the true shape of the angled face A is not seen in the plan or elevation. A view perpendicular to the face from above is necessary. This view projected onto an inclined plane parallel to the face A is an auxiliary plan.

To draw an auxiliary plan of the solid shown in Panel No. 1.
Draw this exercise following, stage by stage, the instructions and the drawings in the panels.

Panel No:
2 Draw the plan and elevation of the solid to the dimensions given adding an extra XY lines in the position shown parallel to the angled face in the elevation. The change of direction line CD bisects the angle between the two XY lines and in this example is at 60°.

3 Project the corners of the elevation at right angles to X_1Y_1.

4 Transfer the width 'w' of the solid into the auxiliary view by projecting parallel to the XY lines.

5 The auxiliary plan is lined in and the true shape of face A is now part of the auxiliary view.

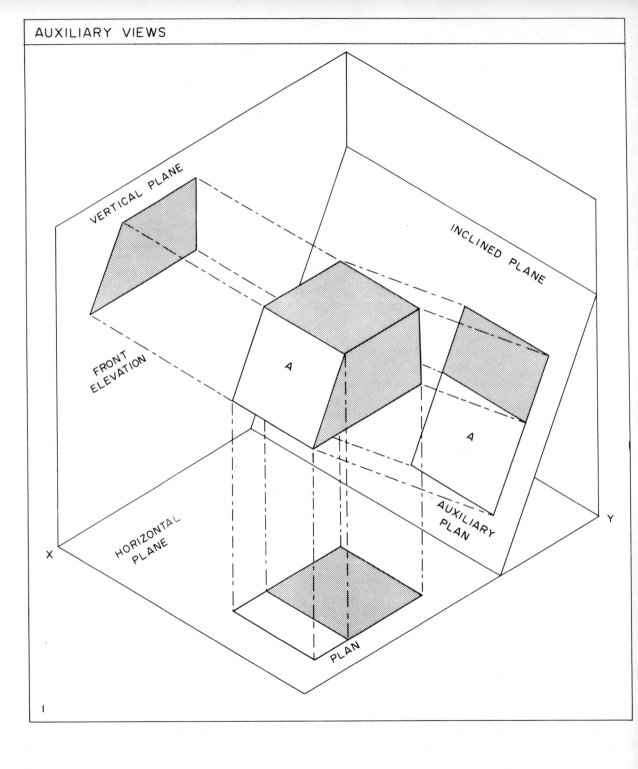

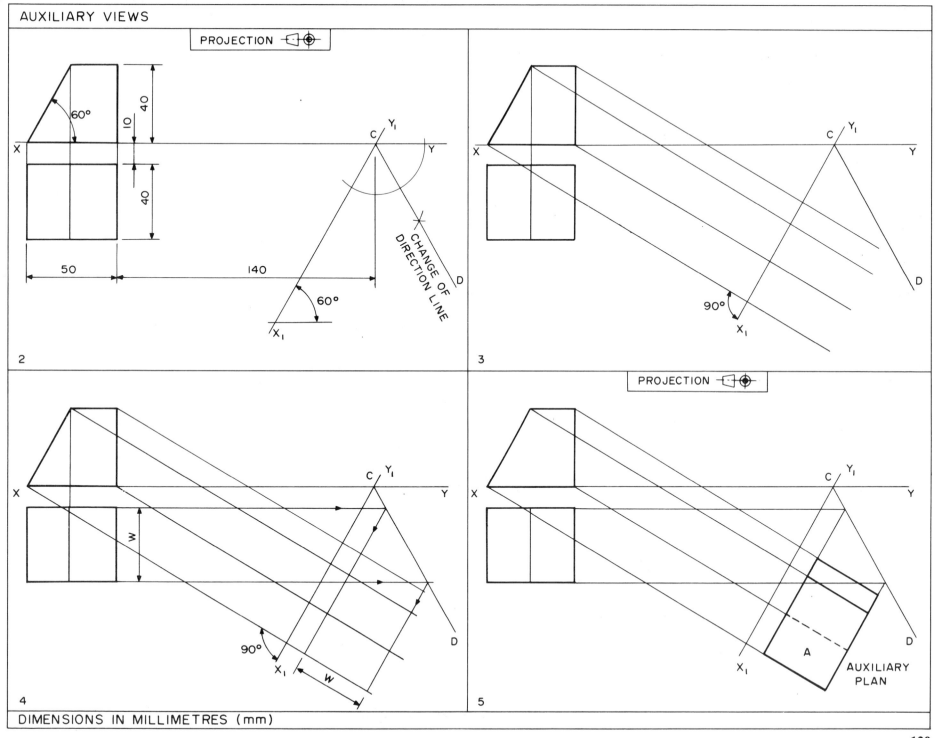

To draw an auxiliary elevation of a block with a curved surface.

Draw this exercise following, stage by stage, the instructions and the drawings in the panels.

Panel No:

1. Draw the shaped block to the dimensions given including the extra XY line and change of direction line CD.

2. Project from the plan, at right angles to the X_1Y_1 line, the corners and the points of tangency of the radius 'a' and 'h'.

    Using the change of direction project the height and centre line of the elevation into the auxiliary view. Mark 'a' and 'h'.

3. Mark all the projected points to establish the limits of the curve in the auxiliary view. Divide the semicircle in the elevation into six equal parts to give further points between 'a' and 'g', and 'h' and 'p'. Project these points into the plan.

4. Project the extra points from plan into the auxiliary view.

5. Using the change of direction line project the extra points into the auxiliary view. Four of these are lettered to make the intersection clear.

6. Line in the auxiliary view joining the points on the projected curve with a freehand line.

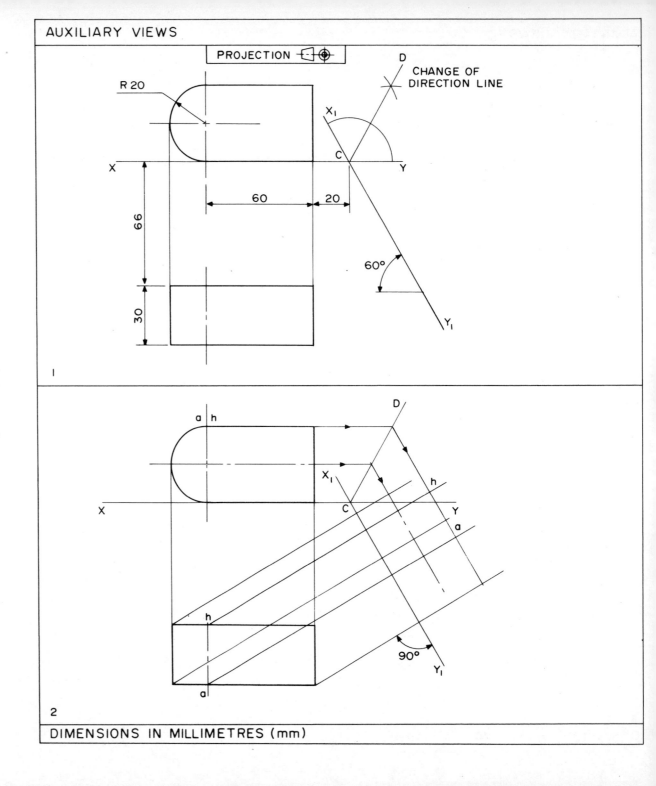

DIMENSIONS IN MILLIMETRES (mm)

AUXILIARY VIEWS

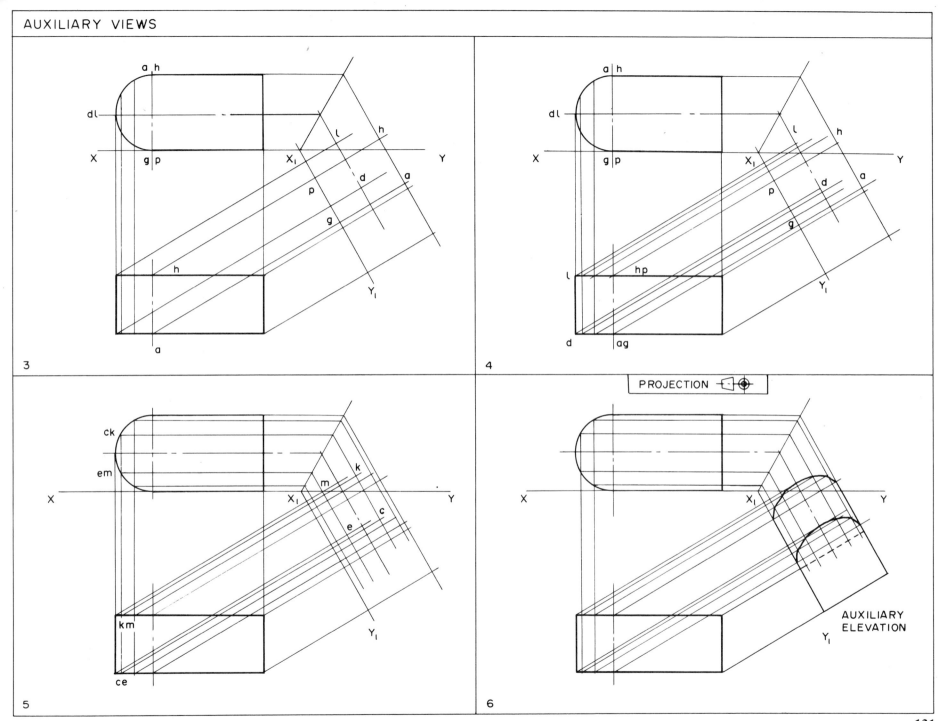

AUXILIARY ELEVATION

To draw an auxiliary plan of a disc.
Draw this exercise following, stage by stage, the instructions and the drawings in the panels.

Panel No:
1. Draw the circular disc to the given dimensions and include the XY lines and change of direction line CD as shown.

2. Divide the circumference of the plan into twelve equal parts and project into the elevation.

3. Project the points on the upper surface of the elevation into the auxiliary view at right angles to the X_1Y_1 line.

4. Using the change of direction line project the points on the plan into the auxiliary view marking the intersection as shown.

5. Join the points with a freehand curve to complete the upper surface of the disc. The thickness of the disc is uniform and to reduce the number of projectors the apparent thickness in the auxiliary plan is found by one projector. This thickness 'T' is marked from each point along the projectors from the plan.

6. Join the lower points with a freehand curve and complete the plan with two lines at 60° touching both curves.

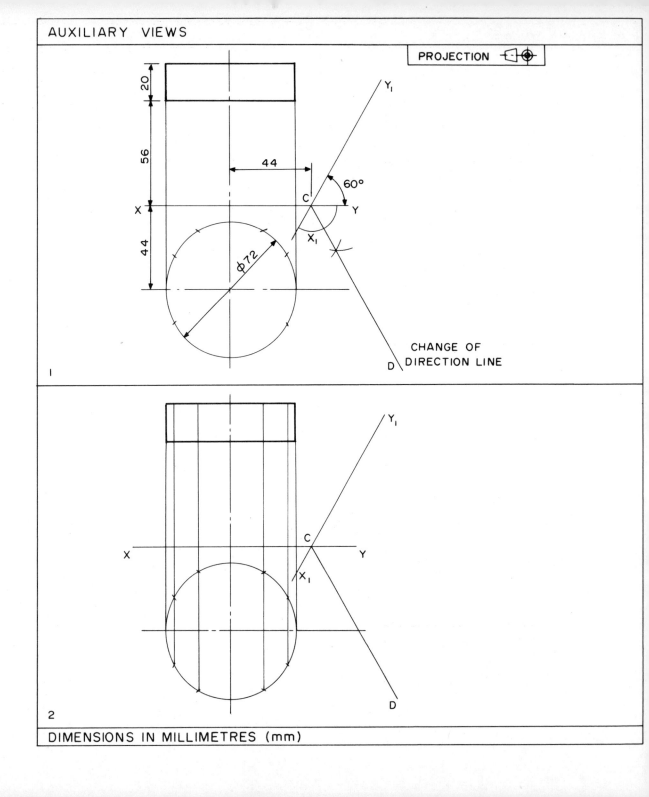

AUXILIARY VIEWS

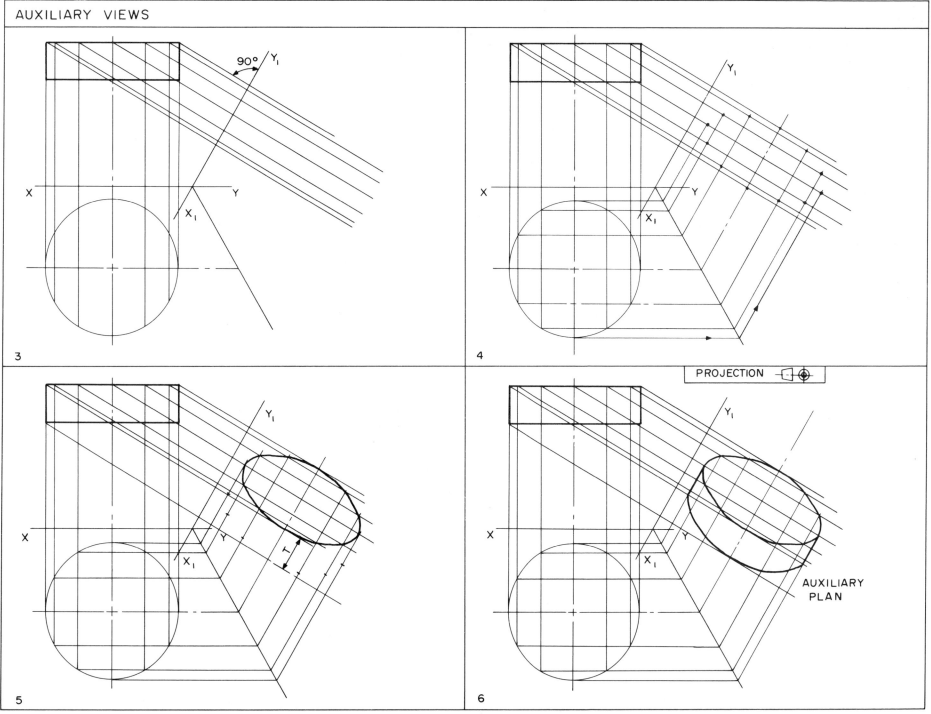

When a solid is cut by an inclined plane the true shape of the section can only be seen by a view perpendicular to the section. This is an auxiliary view and frequently only the surface of the solid contained in the cutting plane is drawn.

To find the true shape of the surface of a square pyramid cut by an inclined plane.
Draw this exercise following, stage by stage, the instructions and the drawings in the panels.

Panel No:
1. Draw the plan and elevation of the given square pyramid cut by a plane at 45° as shown and including the XY line.
2. Draw the second XY line and change of direction line CD as shown and project from the elevation at right angles to X_1Y_1.
3. Project from the plan into the auxiliary view using the change of direction line as shown.
4. Join the projected points in the auxiliary plan to complete the true shape contained by the inclined plane.

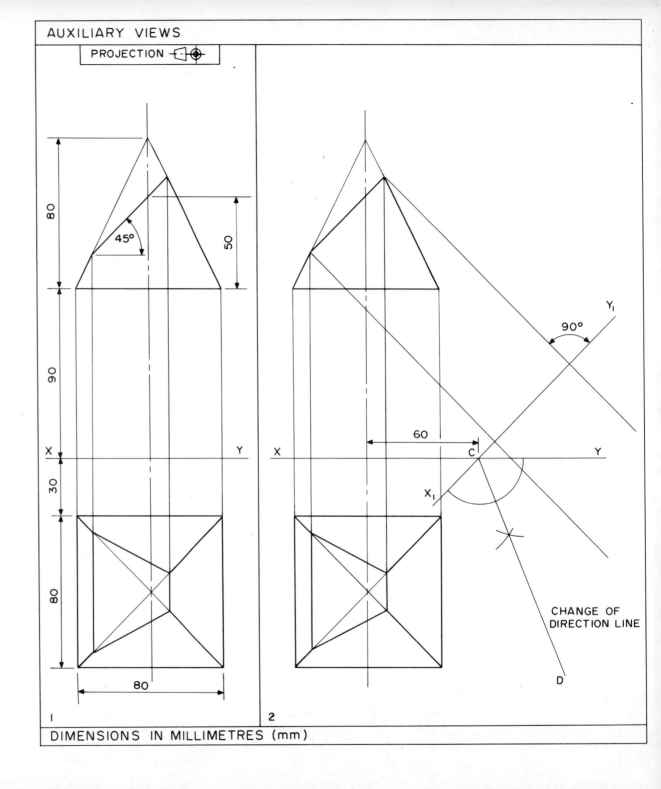

AUXILIARY VIEWS

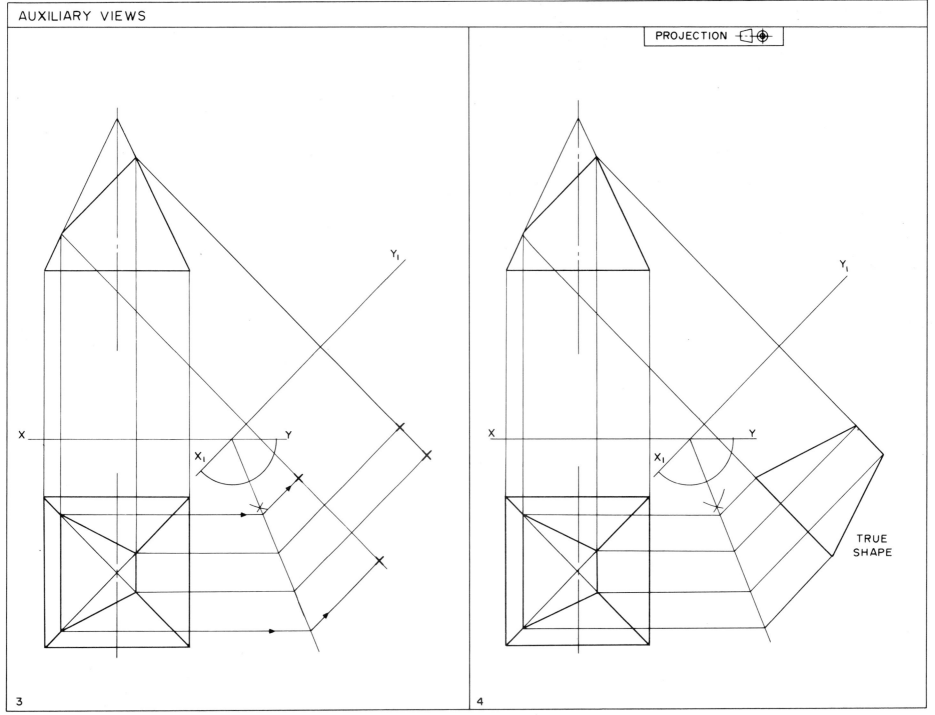

TRUE SHAPE

ISOMETRIC PROJECTION

This is a popular and convenient form of pictorial drawing used for small engineering or other details.

Panel No. 1 shows the plan and elevation of a rectangular prism (a prism is a solid of uniform cross-section). Panel No. 2 shows the same prism in isometric projection. Note the relative positions of corners ABDL and M.

The following points should be memorised:

a) lines and edges e.g. AB and LM which are horizontal in orthographic projection are drawn at an angle of 30° in isometric projection.

b) lines and edges e.g. AD which are vertical in orthographic projection remain vertical in isometric projection.

c) lines and edges which are parallel in orthographic projection are parallel in isometric projection.

Draw in isometric projection the rectangular prism shown in Panel No. 1 following stage by stage the instructions and the drawings in the panels.

Panel No:

4 Draw a vertical line and two lines at 30° to intersect at A as shown (these are often referred to as the isometric axes).

5 Mark points B and C and through them draw vertical lines.

6 Mark point D and through D draw lines at 30° fixing points E and F.

7 Through E and F draw lines at 30° fixing point G.

8 Now line in the complete drawing.

9 Hidden edges can be shown if required.

10-12 These panels show the same prism from different viewpoints.

Exercise:
Draw a rectangular prism 85 mm by 55 mm by 30 mm in each position shown in panels 9, 10, 11, 12.

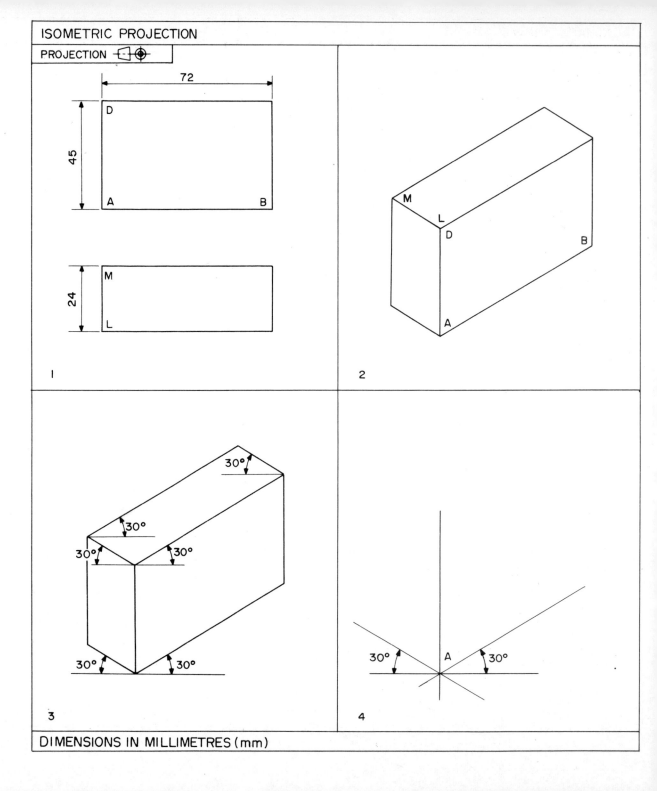

ISOMETRIC PROJECTION

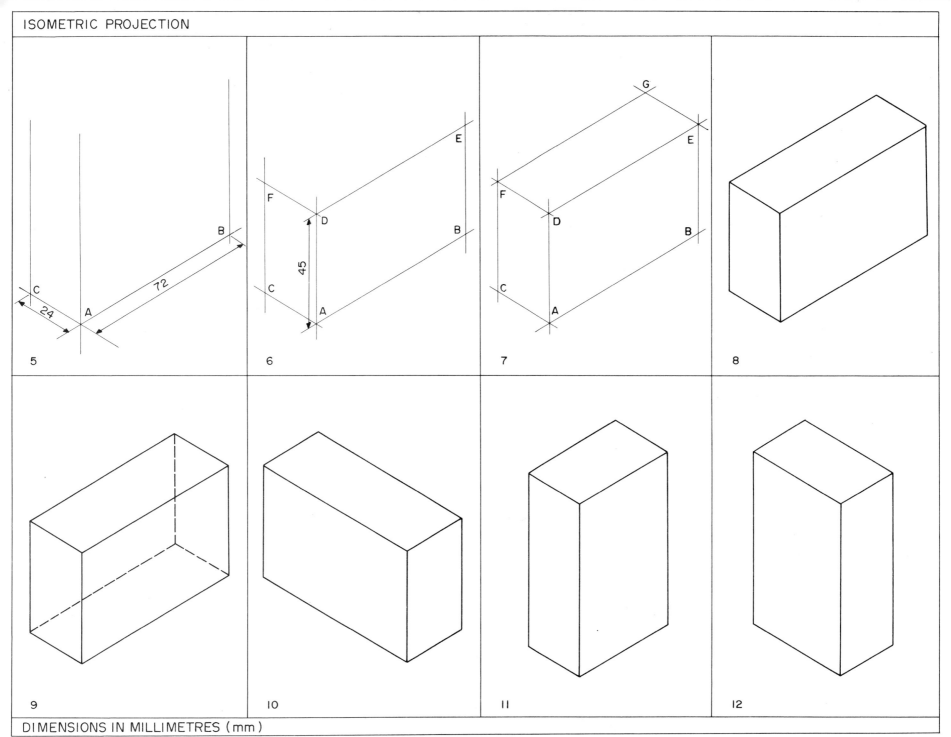

DIMENSIONS IN MILLIMETRES (mm)

Panel No. 1 shows the plan and elevation of a piece of timber prepared for a halving joint.

In Panel No. 2 there is a freehand pictorial drawing.

Before attempting to make an isometric projection you should always make a freehand pictorial drawing of the object.

To make an isometric projection of the halving joint in Panel No. 1.
Draw this exercise following, stage by stage, the instructions and the drawings in the panels.

Panel No:

3 Draw the three isometric axes as shown.

4 Draw an isometric projection of the smallest prism from which the object could be taken, in this case 100 mm long, 50 mm wide and 30 mm deep.
(Note: Just as in carving where one starts with a block and cuts pieces away so in isometric drawing we start with a rectangular or square prism and 'cut pieces away'.

You should not attempt to make an isometric drawing by 'sticking pieces together').

5 From A measure 60 mm along line AB and draw line CD.

6 Through C and D draw vertical lines DE and CF.

7 From D measure 15 mm along line DE and draw line GH.

8 From G and H draw lines GJ and HK.

9 Join points J and K.

10 Now line in the required drawing.

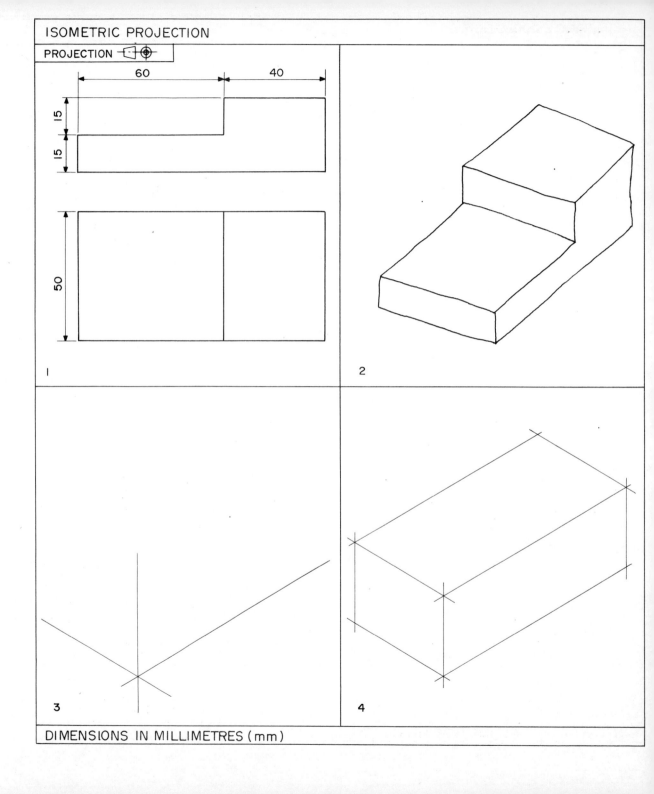

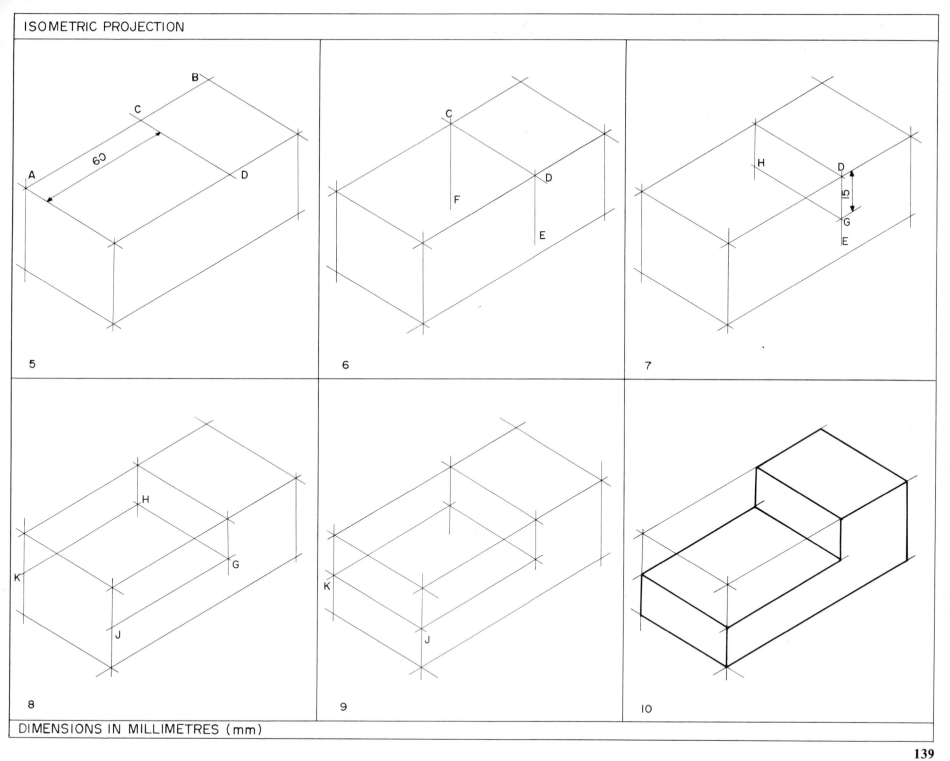

Two elevations of a tenon are shown in Panel No. 1 and a freehand pictorial drawing is shown in Panel No. 2. Three of the corners have been shaded in each panel to help you to relate the drawings to each other.

In the pictorial drawing the tenon has not been viewed to the best advantage, this view has been taken to show you how to deal with the far top corner.

To make an isometric projection of the tenon shown in Panel No. 1.
Draw this exercise following, stage by stage, the instructions and the drawings in the panels.

Panel No:
3 Draw an isometric projection of the smallest prism from which the object could be taken (the dimensions are the same as in the previous drawing).

4 From A measure 60 mm along AB and draw line CD.

5 Draw the vertical line DE and divide it into three equal portions.

6 From F and G draw lines FH and HJ.

7 From H draw line HK.

8 From B draw a vertical line to fix point K on line HK.

9 From K draw a line parallel to FH.

10 Now line in your drawing.

All the hidden edges have been shown in this drawing to show you how this is dealt with. You should not insert hidden detail unless your teacher tells you to do so.

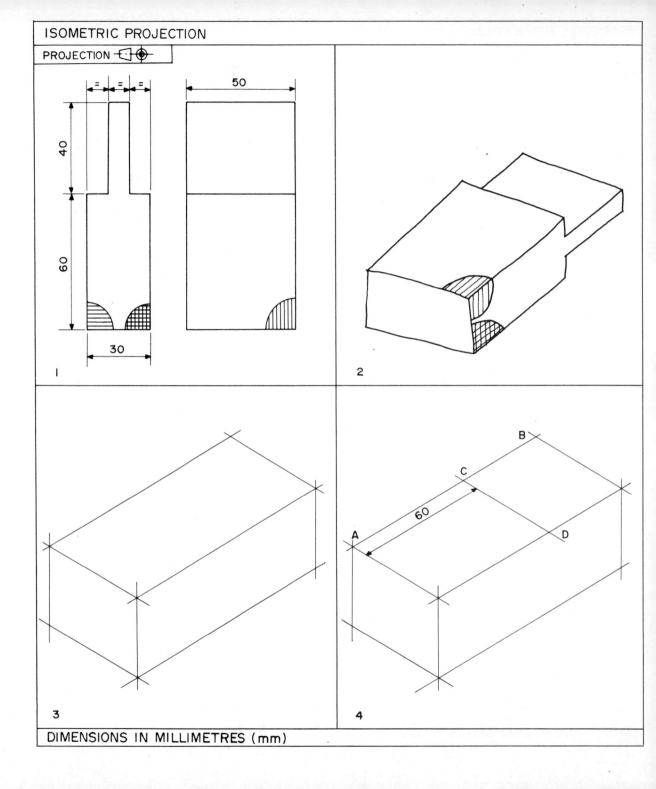

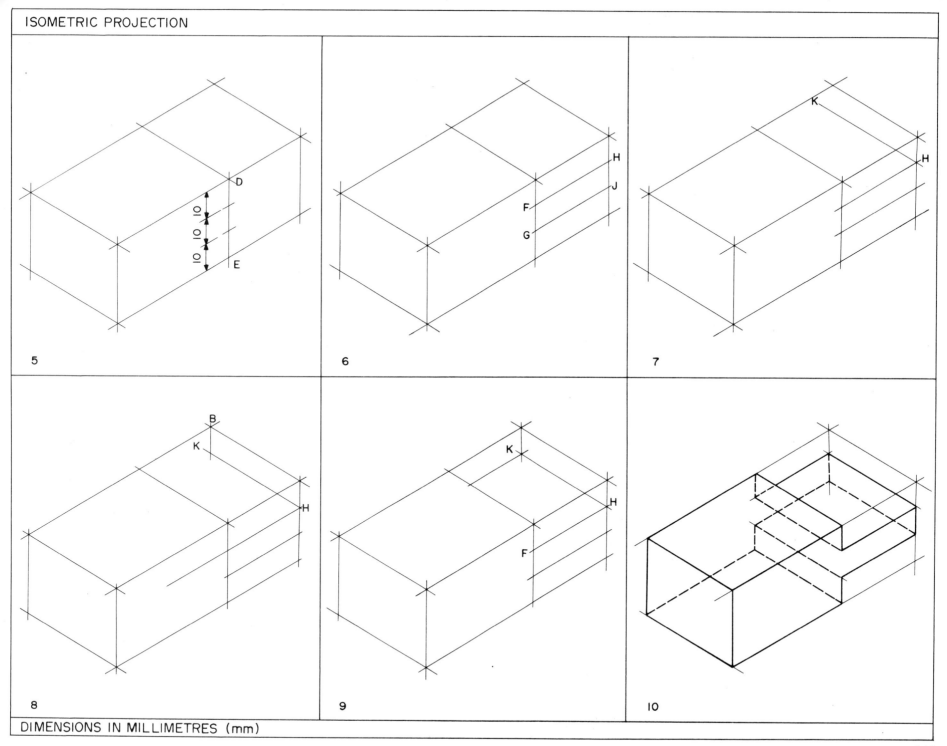

Draw in Isometric Projection, the objects shown in Panels Nos. 1, 3 and 4.

For Panel No. 1 the freehand pictorial drawing has been made for you. Before you start the other two exercises make a freehand pictorial drawing.

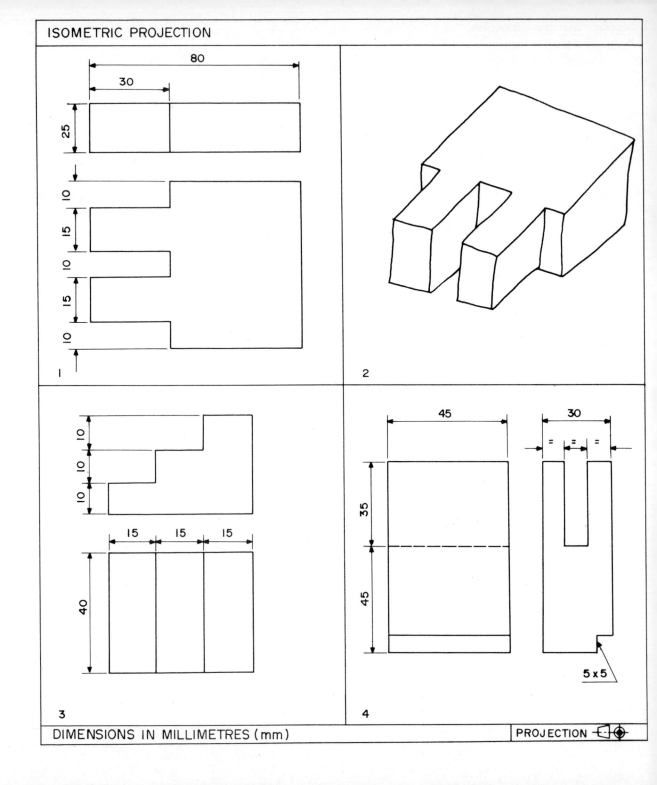

ISOMETRIC PROJECTION

DIMENSIONS IN MILLIMETRES (mm)

Panel No. 1 shows two views of a block, the elevation of which is a square.

Draw this and insert the diagonals AC and BD.

Panel No. 2 shows the isometric projection of the same block. Draw this and insert the diagonals AC and BD as shown.

Now it is obvious that the two diagonals are equal in the elevation (and of course in the actual block) but if you measure the diagonals in your isometric drawing, you will find that AC is shorter than BD. You will also notice that although the diagonals in the elevation form angles of 45° at the corners, they do not form the same angles in the isometric drawing.

This brings out two important points you should remember!

a) Measurements in isometric projection can only be made along the isometric axes (AB, CD and EF in Panel 3) or parallel to them.

  You cannot measure with the rule in positions such as that in Panel 4.

b) You should not attempt to measure angles in isometric projection. As we have already seen angles of 45° and 90° in an elevation do not necessarily measure 45° or 90° in an isometric drawing.

The method of transferring angles from an orthographic projection to an isometric projection will be dealt with later.

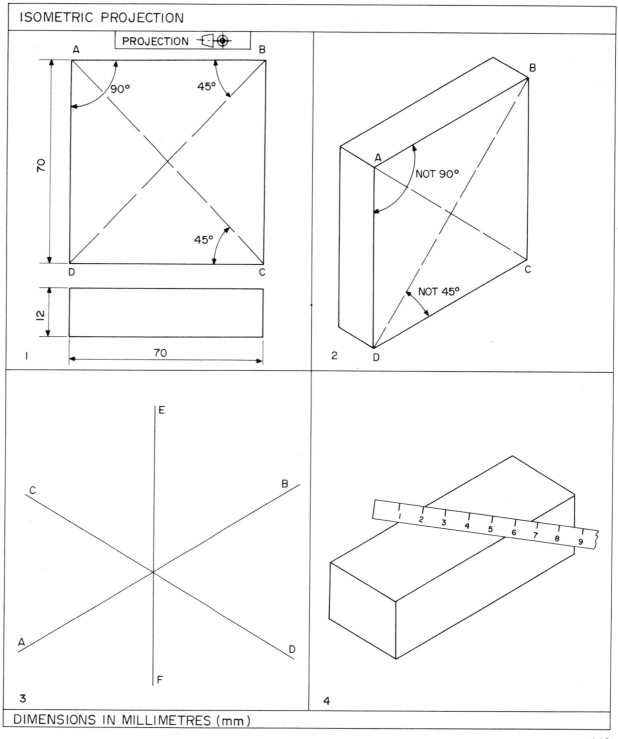

To make an isometric projection of the block shown in Panel No. 1.

Draw this exercise following, stage by stage, the instructions and the drawings in the panels.

Panel No:

2 The freehand pictorial drawing of the block is already drawn for you.

3 If you examine the plan and elevation in Panel No. 1 you will notice that no overall height is given.

Before the 'containing prism' can be drawn, therefore, the elevation must be drawn out to obtain the dimension P.

4 The 'containing prism' can now be drawn. Stepping off the dimension P from the view you have drawn.

5 Draw the ridge line AB in the centre of the top face.

6 The height at which the sloping surfaces begin can now be marked at C and D.

7 Now draw line CE.

In order to fix the position of F on line DF a vertical line must be drawn from G.

8 Lines AC, AD, BE and BF can now be drawn.

9 Now line in the required drawing.

Exercise:

Draw in isometric projection the bracket shown in Panel No. 10.

Note that no overall height is given.

The drawing is not to scale.

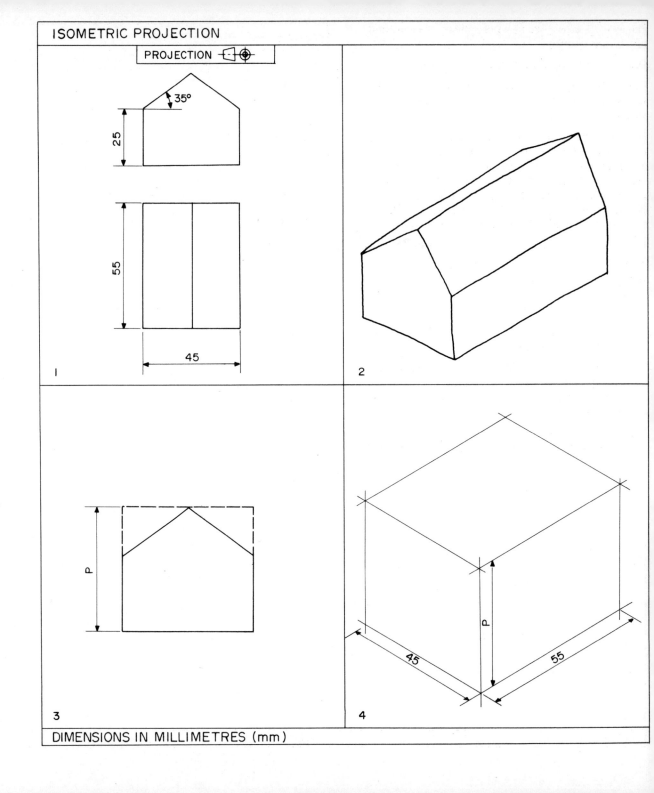

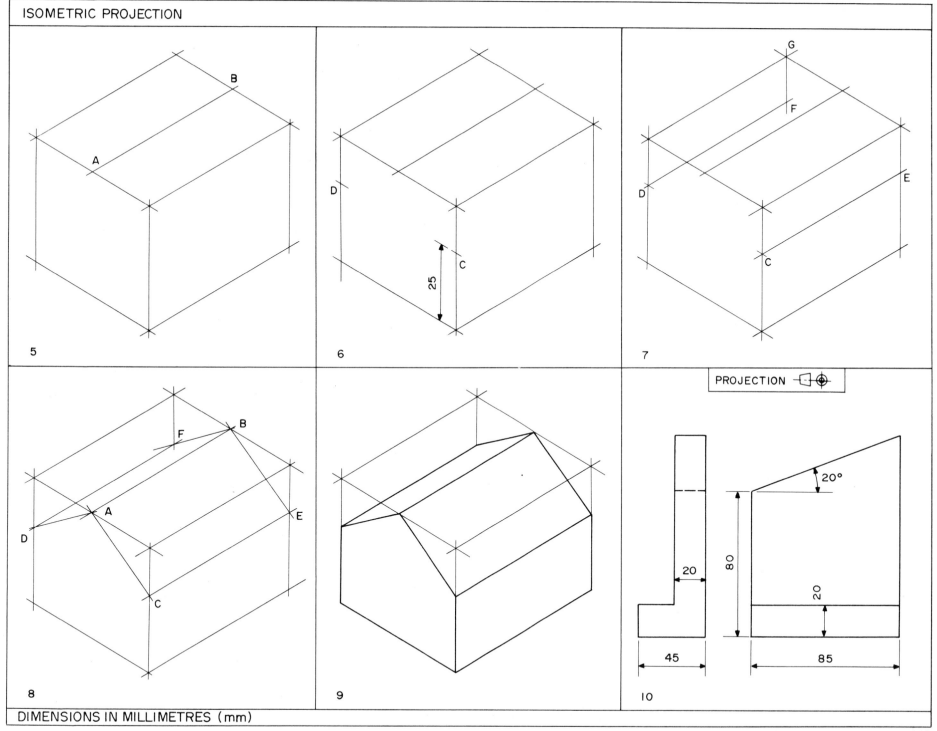

To make an isometric projection of the hexagonal pyramid shown in Panel No. 1.

Draw this exercise following, stage by stage, the instructions and drawings in the panels.

Panel No:

2. The freehand pictorial drawing of the pyramid has been drawn for you.

3. If you examine Panel No. 1 you will notice that only one of the dimensions necessary for drawing the containing prism is given (one of the others will of course be 50 mm). Before the containing prism can be drawn the plan of the pyramid should be drawn and 'boxed in' to obtain the dimensions L and W.

4. The containing prism can now be drawn, stepping off the dimensions L and W from the plan you have drawn.

5. Draw in the hidden lines of the base of your prism. Mark A and B which are in the centre of the shorter sides of the base.

6. Use your dividers to transfer the dimension X from your plan to give you D, C, E and F in the longer sides of the base.

7. Join points A, D, C, B, F and E to form the base of the pyramid.

8. Draw diagonals HJ and KL on the top face of the prism to give you the centre G.

9. Join A, D, C, B, E and F to G.

   Note that lines CB, BE and FE in the base and also lines EG and BG will not be seen.

10. Now line in the required drawing.

    Leave out the hidden detail unless your teacher tells you otherwise.

Exercise:

a) Draw in isometric projection a hexagonal pyramid with a base edge of 30 mm and a vertical height of 75 mm.

b) Draw in isometric projection a hexagonal prism 95 mm long, length of side of hexagon 27 mm.

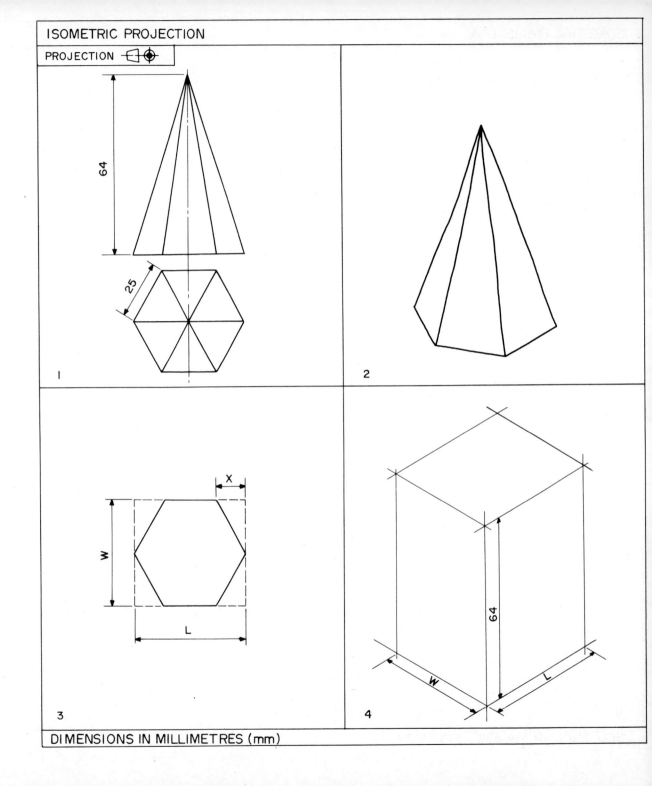

ISOMETRIC PROJECTION

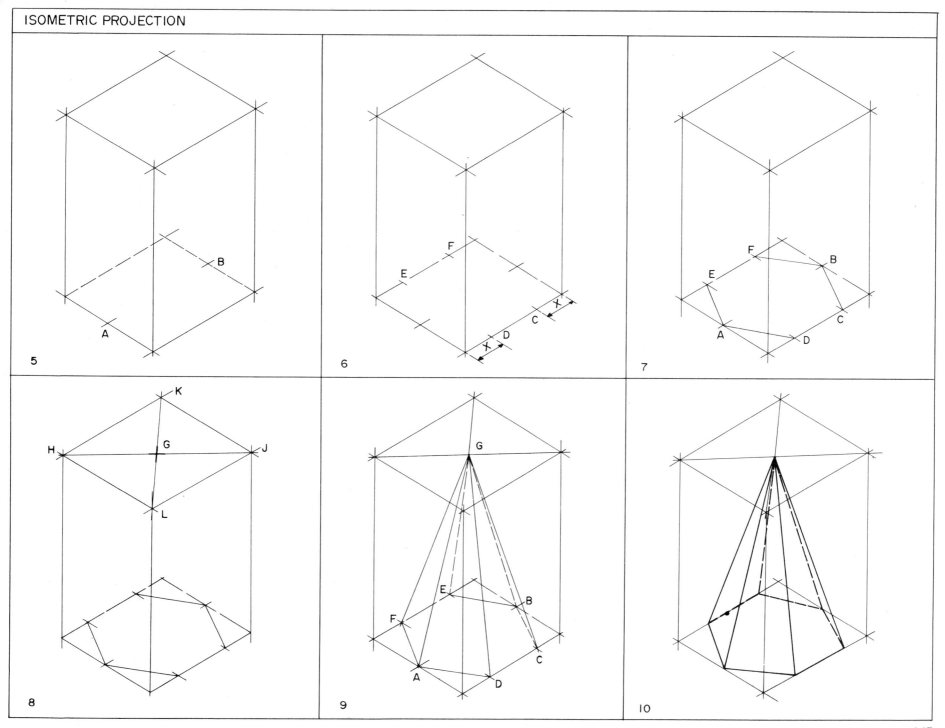

Exercises:

Draw each of the objects in Panels Nos. 1 to 10 in isometric projection.

Panel No. 3 should be drawn five times full size, Panel No. 5 one fifth full size and the remainder full size.

In Panel No. 10 ignore the thickness of the bracket.

All the drawings are in first angle projection and the dimensions are in millimetres.

They are not to scale.

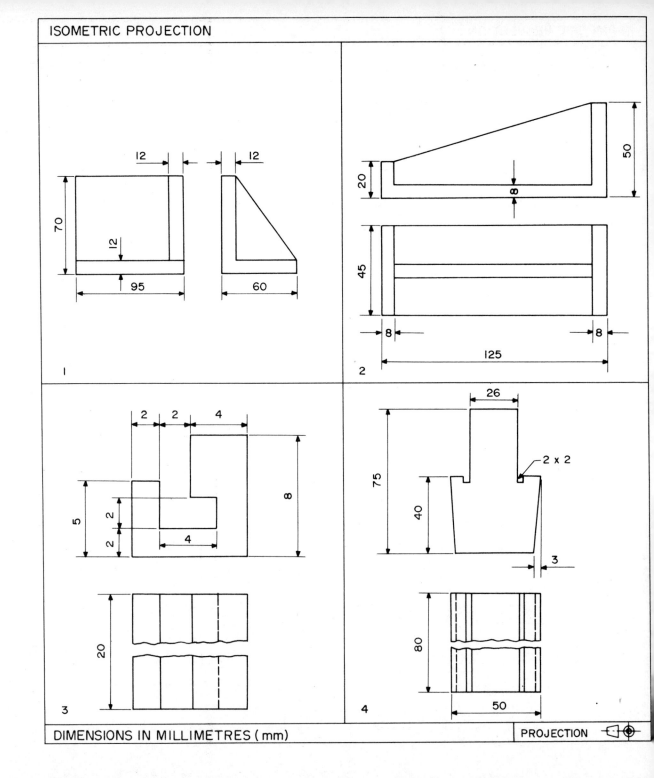

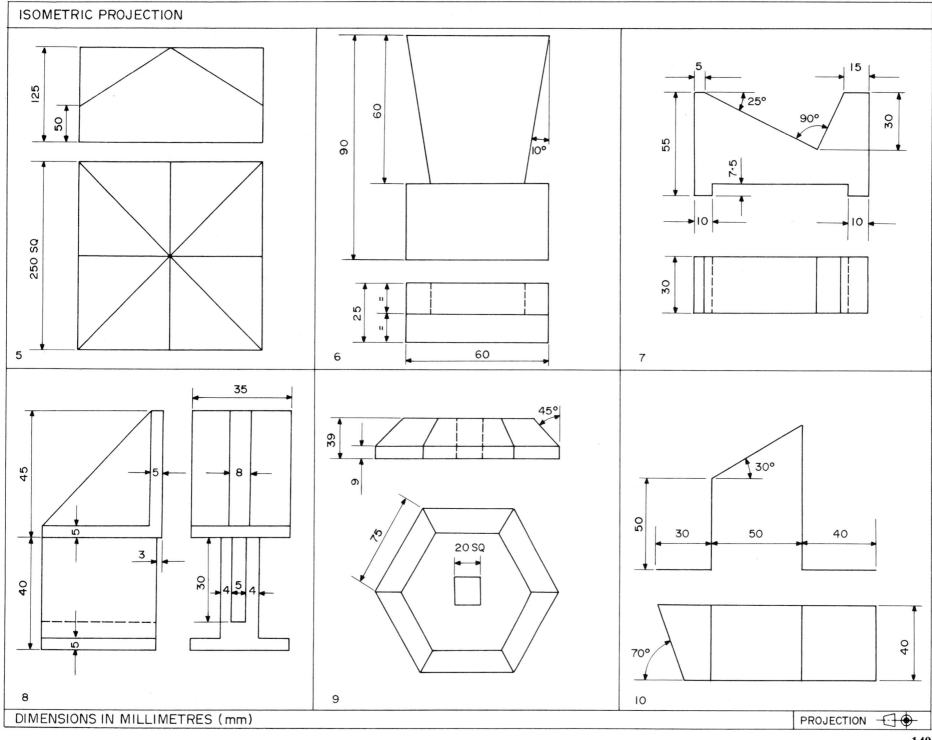

This exercise shows you once again how to deal with angular measurements in isometric projection and also how to deal with measurements which are taken on lines not parallel to the isometric axes (the edges of the containing prism).

You should not attempt this exercise until you have drawn Exercise 1 to 10 starting on page 148.

To draw in Isometric Projection the bracket shown in Panel No. 1.
Draw this exercise following, stage by stage, the instructions in the panels. Do not attempt any stage unless you have understood the previous stages.

Panel No:
1 The plan and elevation of a bracket are shown. Note that neither the overall height nor the overall length is given.

2 This shows a freehand drawing of the bracket.

3 Before the containing prism can be drawn the elevation must be set out, and boxed in as shown, to obtain the length (L) and the height (H). At the same time the positions of three corners and the hole, relative to the sides of the containing 'box' are fixed. (U, W, X, Y and Z).

4 The containing prism can now be drawn. Do not measure L and H but step them off the elevation using dividers.

5 Step off the dimensions U and Z from the elevation and transfer them to the containing prism.

6 Join points Z and U with a straight line.

7 Step off the dimension W from the elevation and transfer this to the containing prism.

8 The line WT can be drawn with your 30° set square as it is parallel to the top back edge of the containing prism.

The line WS can be drawn parallel to ZU.

Join points W and Z with a straight line.

9 In this panel the far end of the containing prism is shown enlarged to almost full size.

The line ZR can be drawn with your 30° set square and to fix the position of point R draw a vertical line downwards from Q.

10 Points R and T can now be joined with a straight line (this is parallel to line WZ).

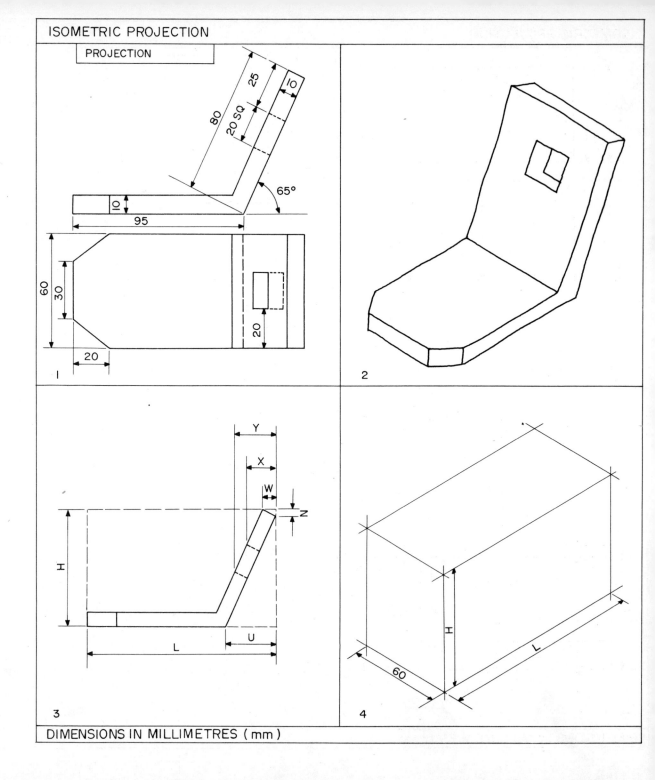

ISOMETRIC PROJECTION

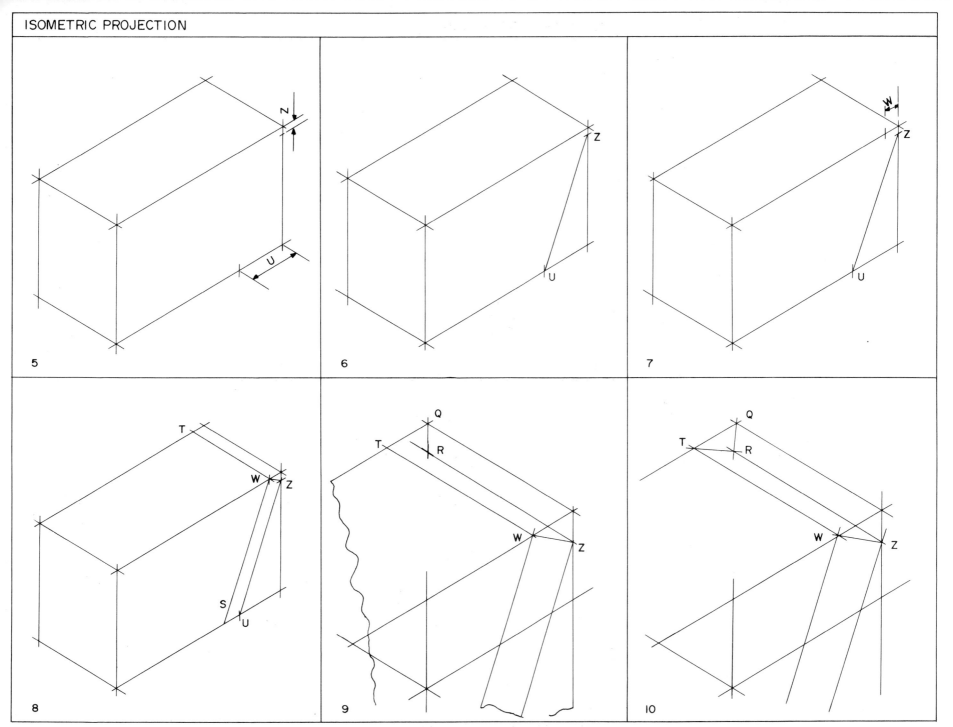

11. Measure vertically 10 mm (the thickness of the bracket) to fix point P.

12. From point P draw lines PS, PP^1 and P^{1}O with your 30° set square.

13. From point S draw a line at 30° to fix point O.

14. Join points T and O with a straight line. (This will be parallel to lines ZU and WS).

15. Measure 15 mm and 20 mm as shown.

16. Draw in the corner details. Note that lines AB are vertical.

17. As the drawing is now complete apart from details of the hole it can be lined in as shown.

18. This and the next six panels show an enlarged view of a portion of the sloping part of the bracket and show the method for fixing the position of the hole.

    Set your dividers to the measurement X in your elevation and mark this from the corner C of the containing prism, along the top near edge to fix point X^1. Draw a line vertically downwards from X^1 to fix point X^2 on the edge of the bracket.

    Draw line X^{2}N with your 30° set square.

19. Now set your dividers to the measurement Y in your elevation and mark this from corner C of the containing prism along the top near edge to fix point Y^1.

    Draw a line vertically downwards from Y^1 to fix point Y^2 on the edge of the bracket.

    Draw line Y^{2}M with your 30° set square.

20. From the edge of the bracket, along line NX^2 measure 20 mm to fix point K and then a further 20 mm to fix point J.

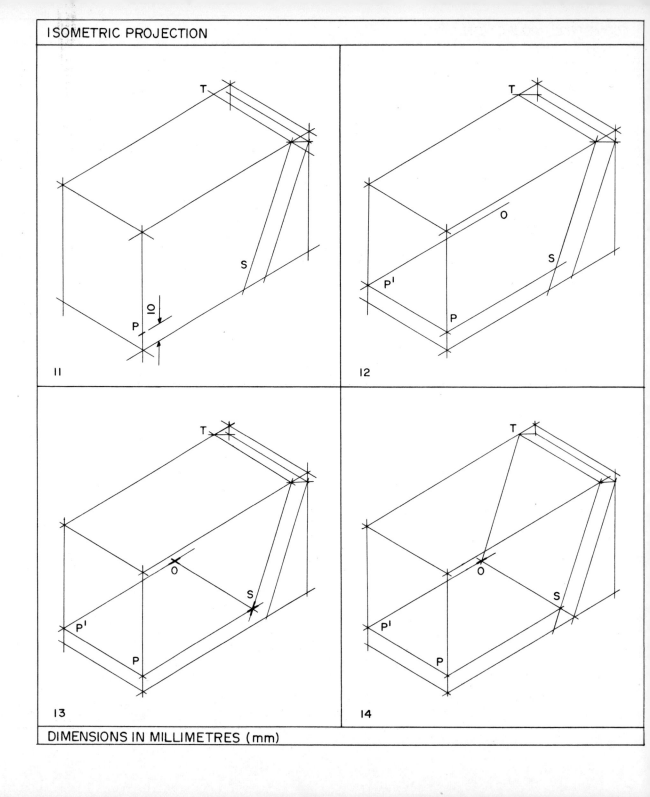

ISOMETRIC PROJECTION

DIMENSIONS IN MILLIMETRES (mm)

ISOMETRIC PROJECTION

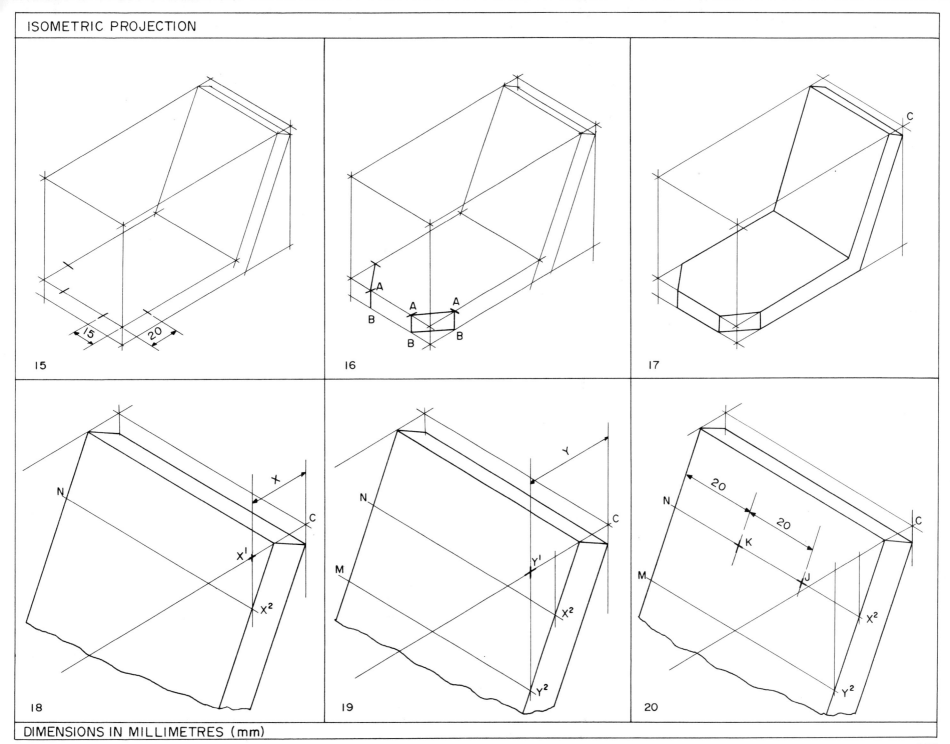

DIMENSIONS IN MILLIMETRES (mm)

21 Draw lines KG and JG^1, both parallel to lines WS, ZU and TO.

22 From point G draw line GD and from point Y^2 draw line Y^{2}E. Both are parallel to line WZ.

23 With your 30° set square draw a line from point E to intersect line GD and fix point D.

24 Draw line DD^1 parallel to line KG.

Now complete the lining in of your drawing.

25 This panel shows the completed drawing lined in.

Exercises:

a) Draw, full size, in isometric projection the block shown in Panel No. 26.

b) Draw, full size, in isometric projection, the strap shown in Panel No. 27.
(Note: Both of these exercises pose problems similar to those in the exercise you have just completed).

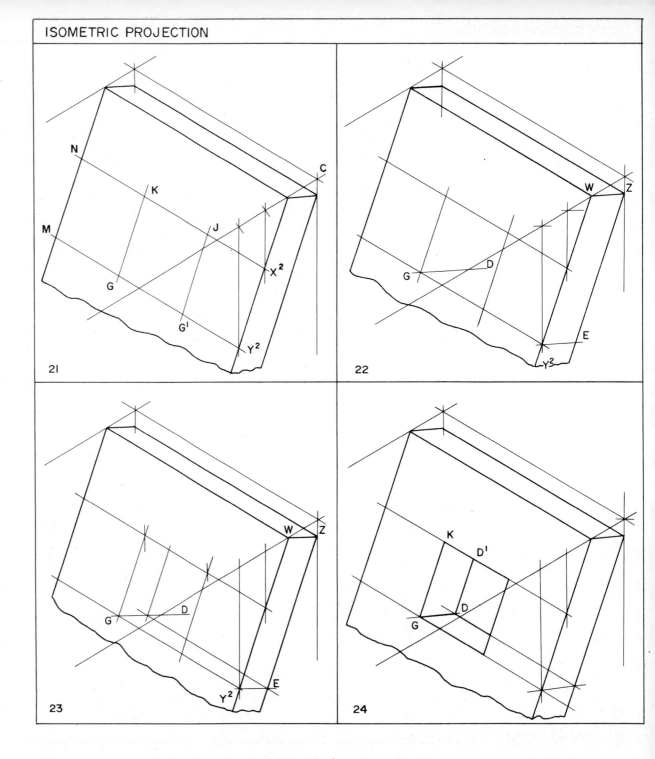

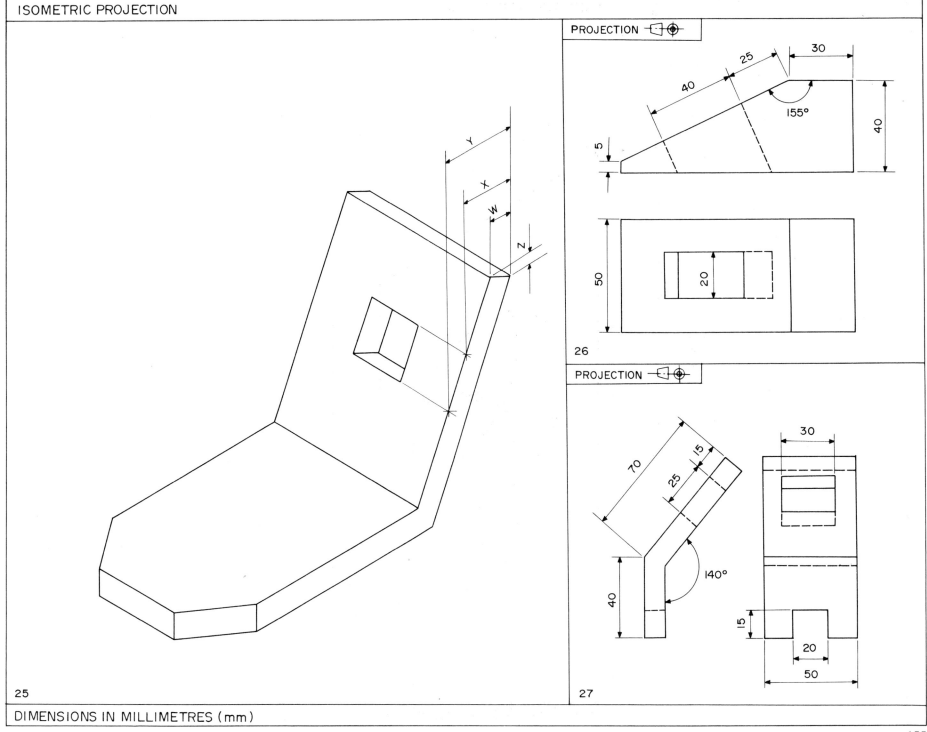

Drawing curves in isometric projection.
Panel No. 1 shows an isometric drawing of a metal plate pierced by a circular hole.

The isometric projection of a circle is an ellipse and the isometric projection of part of a circle will, likewise, be part of an ellipse.

There are several ways of drawing circles in isometric projection and the method used here produces, not a true ellipse, but an approximation. It has the advantage, however, that curves can be drawn with compasses.

To draw the Plate shown in Panel No. 1.
Draw this exercise following, stage by stage, the instructions and the drawings in the panels.

Panel No:
2. Make an isometric drawing, full size, of the plate and draw the centre lines of the hole, EF and GH.
3. Draw the 'square' ABCD with sides equal to the diameter of the hole (50 mm).
4. Draw the diagonals of the 'square' AC and BD.
5. Join points B and F with a straight line. The intersection of line BF with the diagonal AC fixes Point R1 which is one of the centres.

   Join points D and E also with a straight line. The intersection of line DE and the diagonal AC fixes centre R2.

   The corners B and D are the other two centres for the curve.

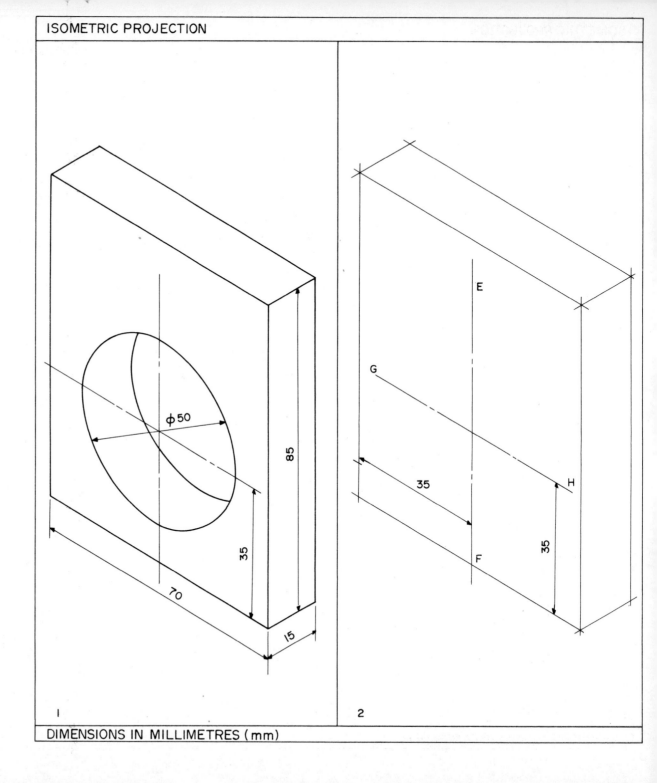

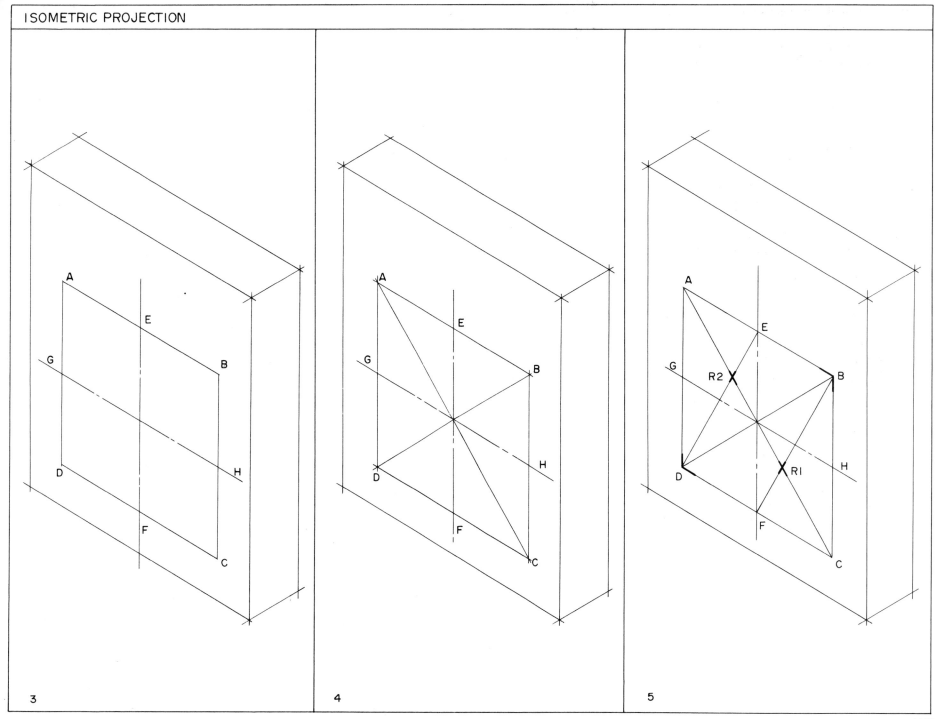

6 With the point of your compasses at B and set to a radius equal to BF, draw the curve from F to G.

   With the point of your compasses at D and set to the same radius, draw the curve from E to H.

7 With the point of your compasses at R1 and set to a radius equal to R1H, draw the curve from H to F.

   With the point of your compasses at R2 and set to the same radius, draw the curve from G to E.

8 Most of the construction lines have been left off the drawing in this panel, but the centres B, D, R1 and R2, from which the curves have been drawn, have been left in.

   If you look back to Panel No. 1, you will see that part of the curve which occurs in the back face of the plate can be seen. This curve is, of course, parallel to that on the front face.

   To find the centres from which it can be drawn, lines are projected at 30° from B, D, R1 and R2. The required centres (B1, D1, R3 and R4) will lie on these lines 15 mm (the thickness of the plate) away from B, D, R1 and R2 respectively.

9 Draw lines at 30° from F and G.

   With the point of your compasses at B1 and set to a radius equal to BF, the curve from F1 to G1 can now be drawn.

   Note that centre D1 is not needed as the portion of the curve drawn from it is unseen.

10 With the point of your compasses first at R3 and then at R4 and set to a radius equal to R1H, the remaining two portions of the curve can be drawn.

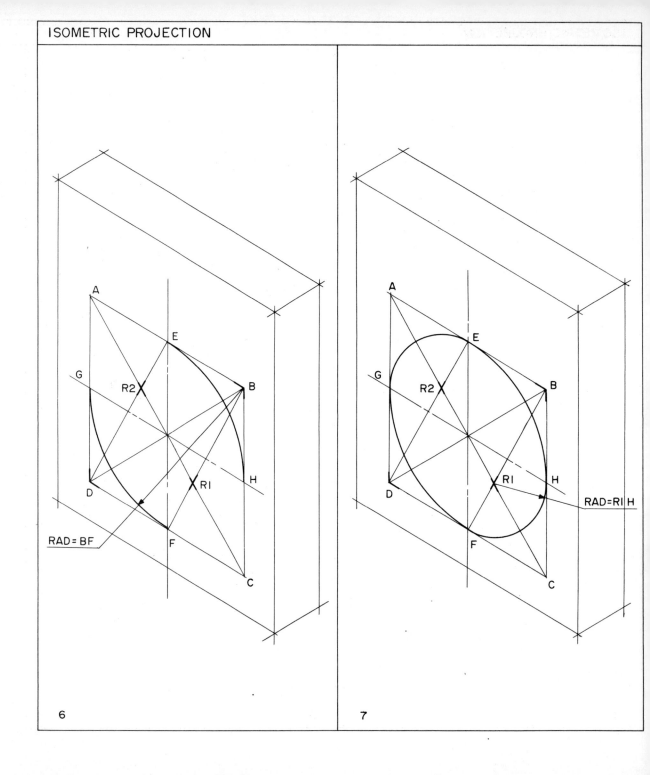

ISOMETRIC PROJECTION

ISOMETRIC PROJECTION

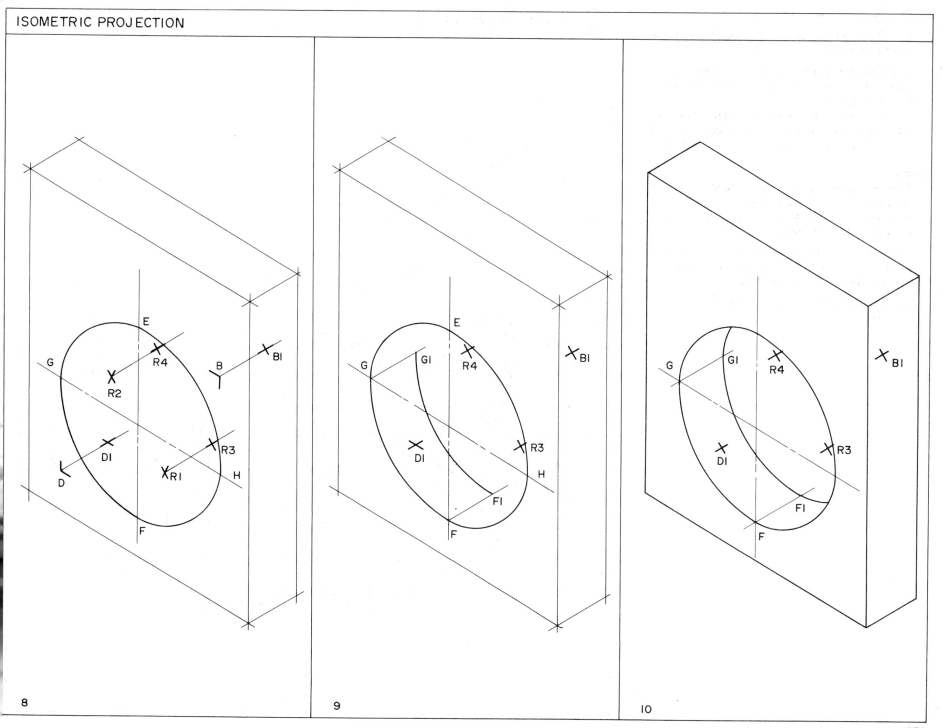

11   The drawing can now be lined in.

This Panel shows the drawing complete with all construction lines.

12,13   If the plate had been drawn as shown in these panels the construction, as can be seen, would have been similar.
(Note: It is useful to remember that two of the four centres are the ends of the shorter diagonal and that the other two lie on the longer diagonal.

Exercises:

A 60 mm cube with a circle inscribed on each of its visible faces is shown in Panel No. 14. The construction for the circle on the top face is shown. Draw the cube, showing the three circles.

It is worth repeating this exercise, varying the size of the cube each time, until the method of construction has been thoroughly mastered.

The drawings in Panels Nos. 12 and 13 should also be copied.

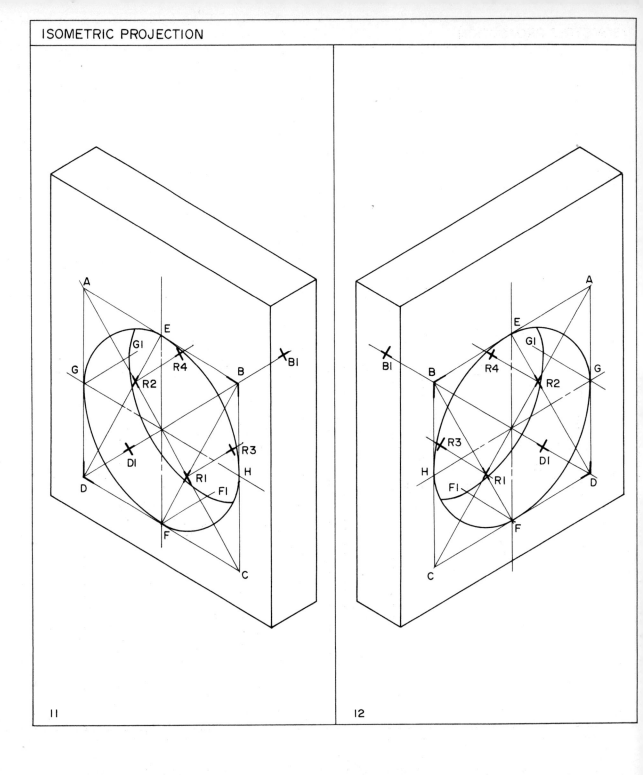

ISOMETRIC PROJECTION

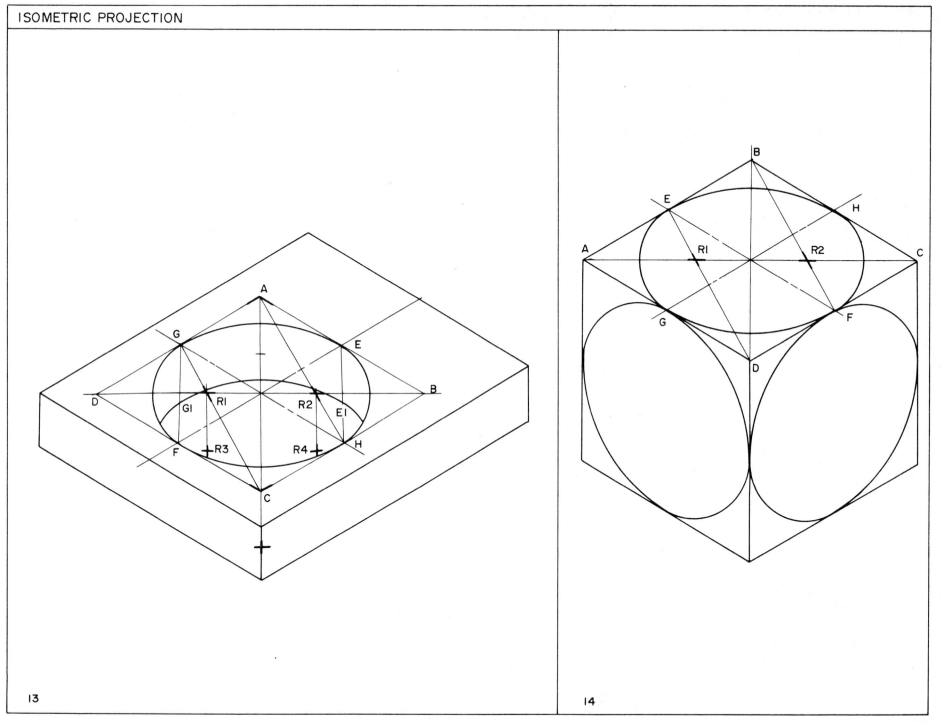

To draw the block shown in Panel No. 1.
Draw this exercise following, stage by stage, the instructions and the drawings in the panels.

Panel No:

1. This shows a plan and elevation of a block with a curved surface. The curve is semi-circular in elevation.

2. This shows an isometric projection of the circle of which the semi-circle ABC forms a part.

3. Draw the containing prism and draw in the straight detail.

4. Draw the 30 mm 'square' and draw in its diagonals GE and HJ and its centre lines AC and BD.

5. The corner of the 'square' E is one of the centres for the curve and, by joining points E and B, the other centre F can be fixed on diagonal HJ.

6. This and Panel No. 7 show an enlarged view of the front curved face. With the point of your compasses at E and set to a radius equal to AE, draw the curve from A to B.

7. With the point of your compasses at F and set to a radius equal to BF draw the curve from B to C.

8. This shows your drawing as it should appear now.

9. This and Panels Nos. 10 and 11 again show an enlarged view of the curved part of the drawing.

    To obtain the centres from which the curve in the back face can be drawn project a line from F at 30° and mark centre F^1 35 mm (the thickness of the block) from F.

    From E along diagonal EG mark centre E1 35 mm from E.

    From A and B project lines at 30° to fix points A^1 and B^1.

10. With the point of your compasses at E^1 and set to a radius equal to AE (or $A^1 E^1$) draw the curve from A^1 to B^1.

    With the point of your compasses at F^1 and set to a radius equal to BF (or $B^1 F^1$) draw the remainder of the curve.

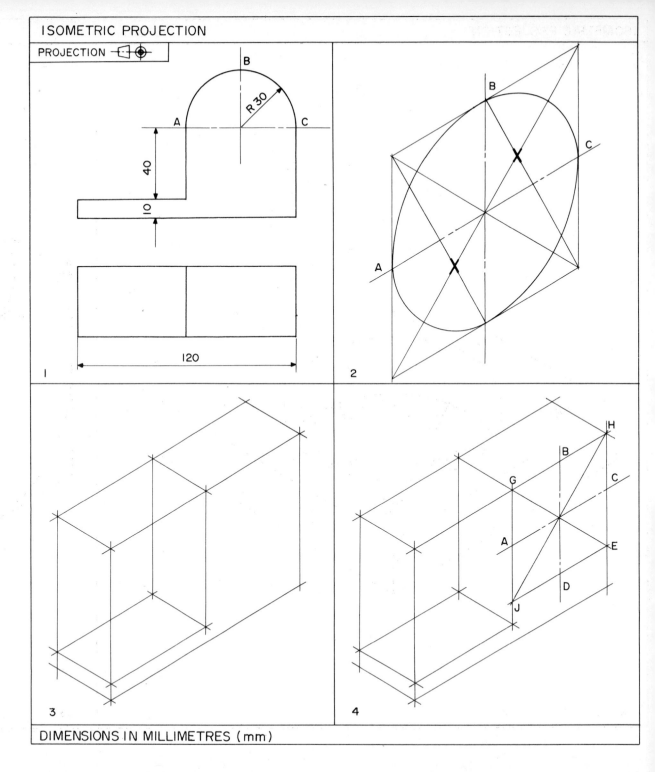

DIMENSIONS IN MILLIMETRES (mm)

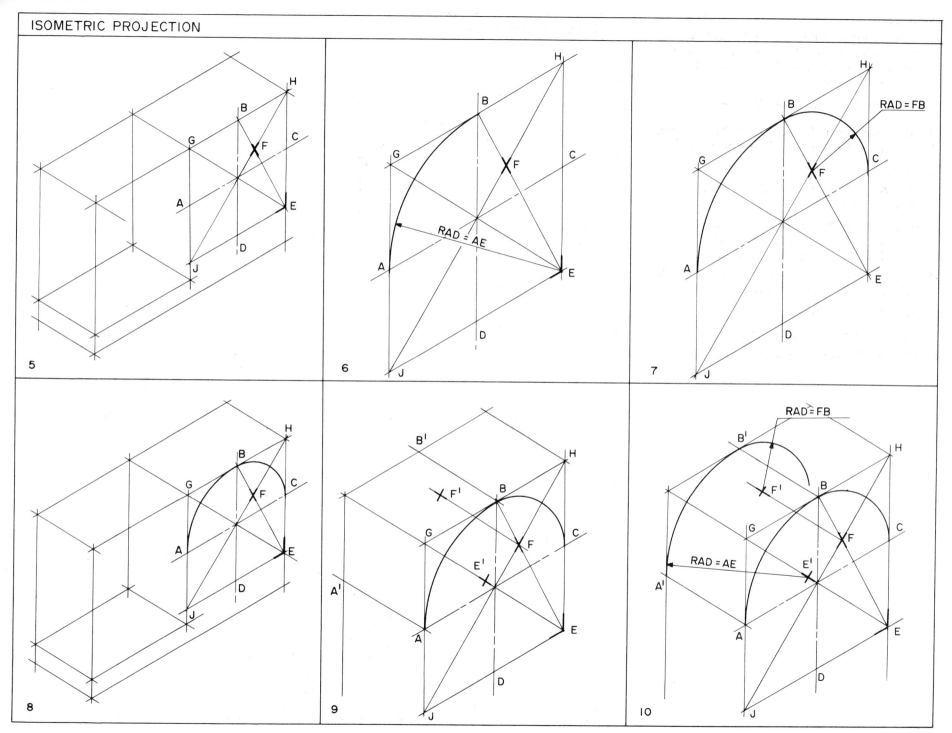

Panel No:

11 Only a portion of the curve in the back face will be seen, and the line TN drawn at 30° and tangential to both curves indicates where the visible portion of the curve ends.

12. This shows the completed drawing lined in.

To draw the object shown in Panel No. 13 using isometric projection making n the lowest corner.

Panel No:

14 This shows the first stages of the drawing with n correctly placed.

If the hole of 20 mm diameter is considered too small to construct in isometric using compasses the following panels show an alternative method.

15 Draw the isometric axes of the ellipse and mark R, the radius of the circle as shown.

16 Mark R on the other axis.

17 At 30° draw through each point as shown.

18 Complete the ellipse freehand.

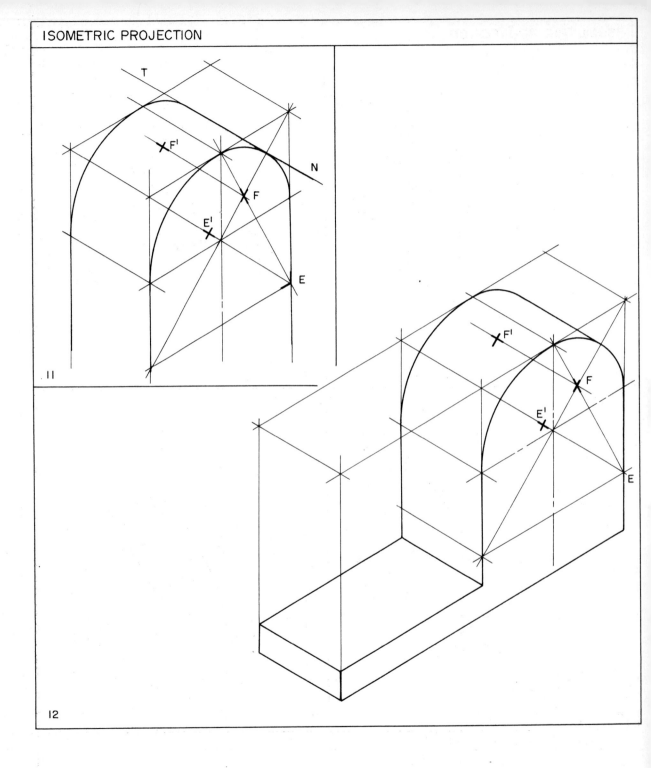

ISOMETRIC PROJECTION

164

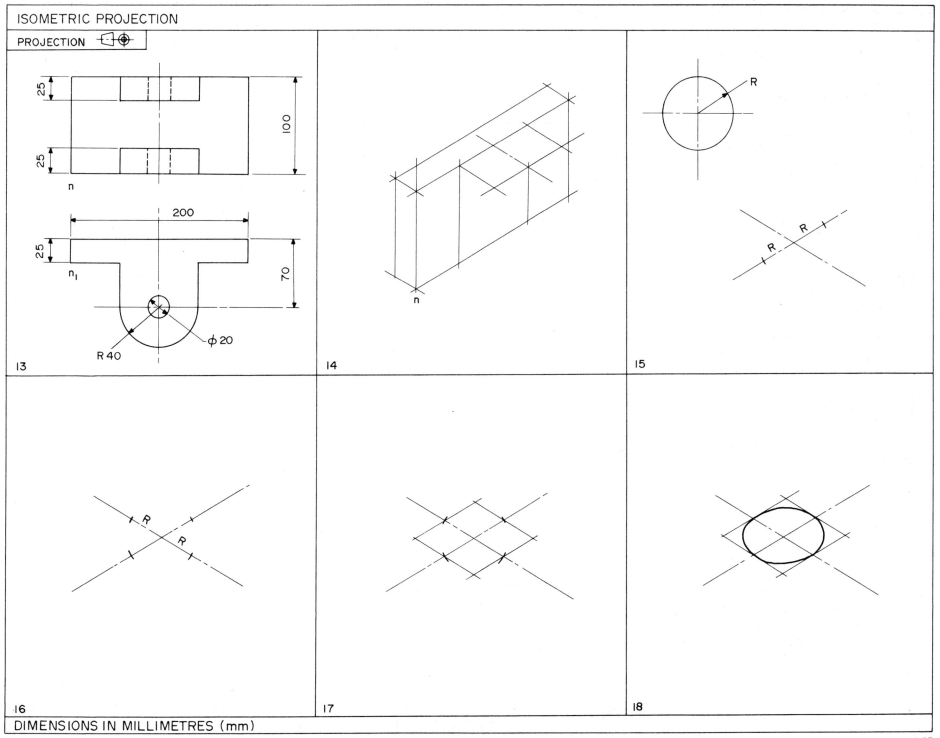

To make an Isometric Projection of the Bracket shown in Panel No. 1.

Draw this exercise following, stage by stage, the instructions and the drawings in the panels.

Panel No:

2 The freehand pictorial drawing of the Bracket is done for you.

3 It is sometimes, as in this case, more convenient to draw two containing prisms. Draw these full size.

To begin with ignore the 36 mm dia. hole and concentrate on drawing the rounded end of the bracket.

Complete the square ABCD, draw in the diagonals and the centre lines EF and GH.

4 As we are only drawing a semi-circle, only two centres will be needed.

One of these is the corner of the 'square' B and the other is obtained by joining points B and E with a straight line. Where this crosses the diagonal AC we have the other centre R.

5 With the point of your compasses on B and set to a radius equal to BH draw the arc from H to E.

With the point of your compasses on R and set to a radius equal to RE draw another arc from E to G.

6 Draw lines vertically downwards from B and R and from each of these points measure 15 mm (the thickness of the bracket) giving you centres B^1 and R^1.

Draw lines vertically downwards from E and H to fix points E^1 and H^1.

7 With the point of your compasses on B^1 and set to a radius equal to B^1H^1 draw the arc from H^1 to E^1.

Only a part of the next arc is seen and to show where it ends draw the vertical line TN tangential to the curve. Now with the point of your compasses on R^1 and set to a radius equal to R^1E^1 draw an arc from E^1 to touch the line TN.

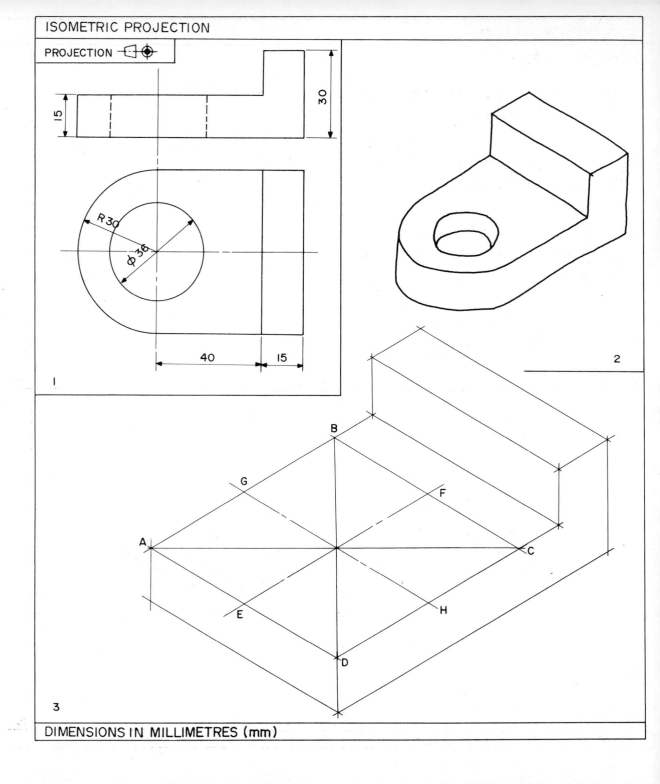

DIMENSIONS IN MILLIMETRES (mm)

ISOMETRIC PROJECTION

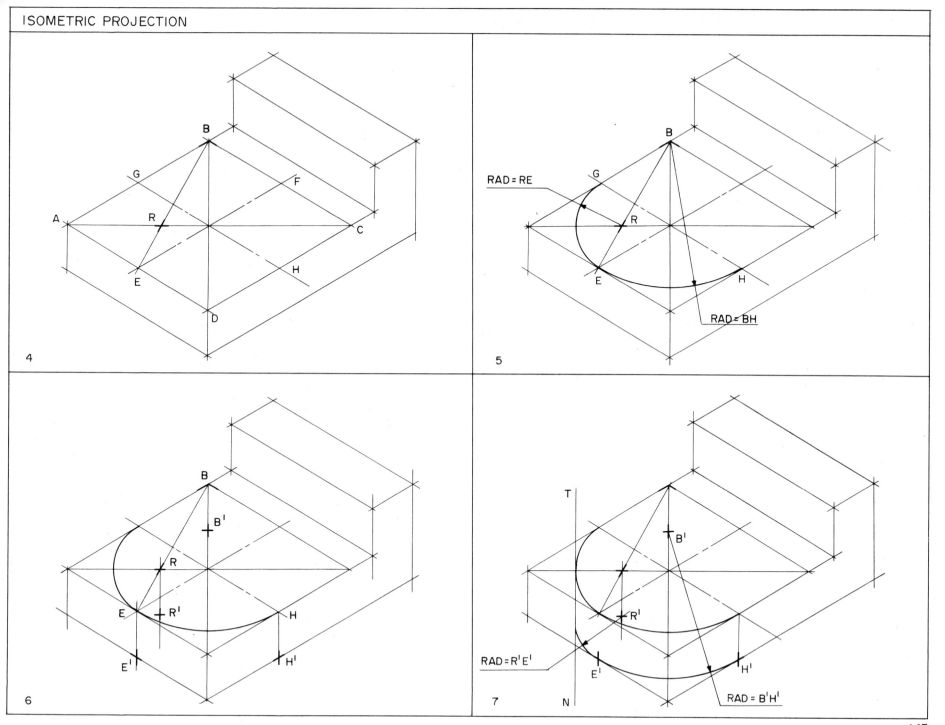

8 Line in your drawing as shown including that portion of the line TN which is part of the drawing.

9 In the next few panels a lot of the construction lines which are no longer needed have been left out. We now have to deal with the 36 mm dia. holes.

Draw the 'square' JKLM with 36 mm sides. The diagonals and centre lines will already be on your drawing.

The corners K and M will be two of the centres and by joining K to E2 and F1 to M with straight lines the other two centres R2 and R3 are fixed on the diagonal JL.

10 With the point of your compasses on M and set to a radius equal to MF1 draw the arc from G1 to F1.

With the point of your compasses on K draw the arc from H2 to E2.

11 With the point of your compasses on R2 and set to a radius equal to R2 E2 draw the arc from E2 to G1.

With the point of your compasses on R3 draw the arc from F1 to H2.

12 Only part of the circle on the underneath face is seen and this can now be drawn.

Measure 15 mm vertically downwards from M to fix centre M1.

Draw lines vertically downwards from G1 to F1

With the point of your compasses on M1 and set to a radius equal to MF1 draw an arc from G2 to F2.

13 Draw lines vertically downwards from R2 and R3 and measure 15 mm along these to fix centres R4 and R5.

With the point of your compasses on R4 and set to a radius equal to R2 E2 draw the small arc from G2.

With the point of your compasses on R5 draw the small arc from F2.

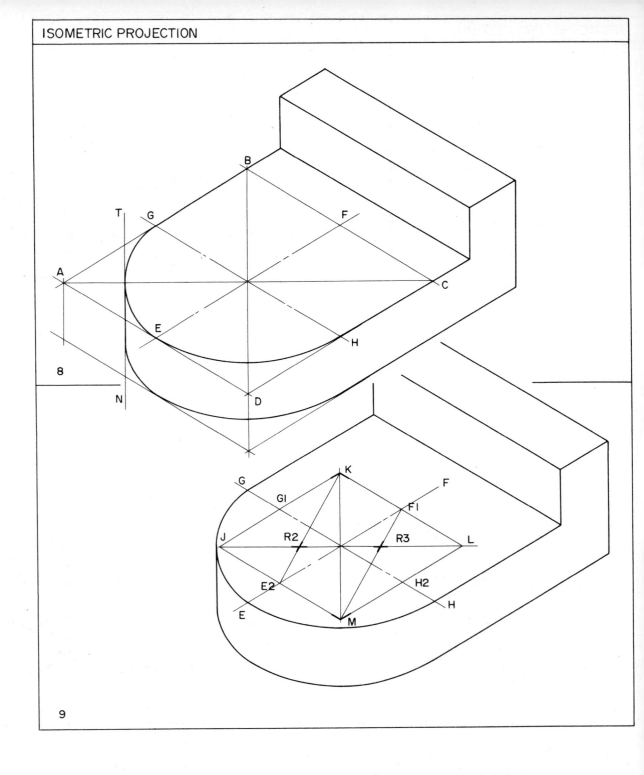

ISOMETRIC PROJECTION

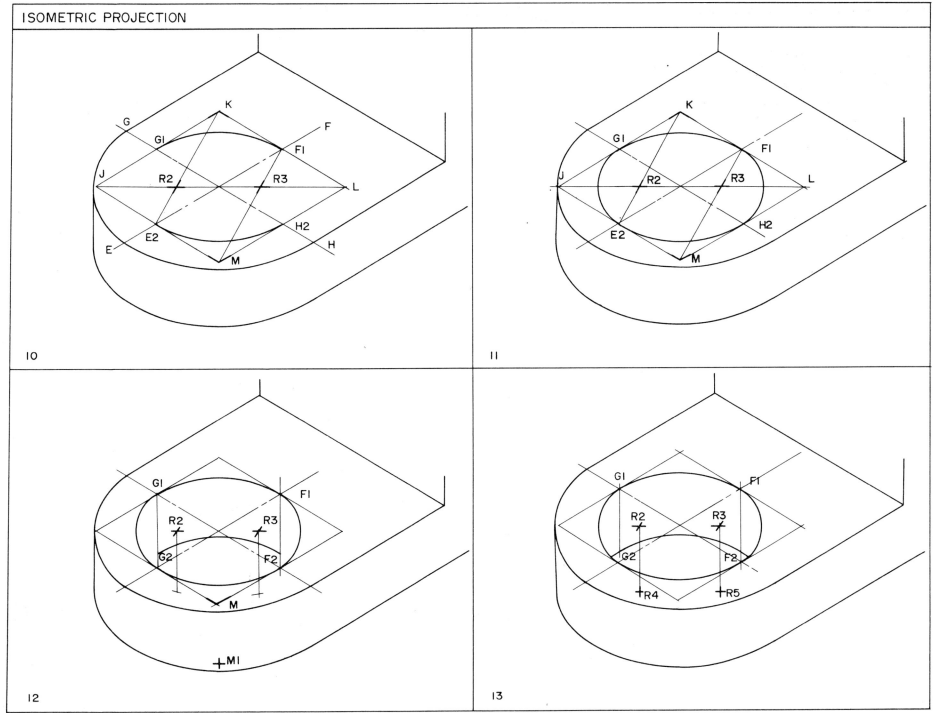

14 This panel shows the completed drawing.

Exercises:

Draw, full size, in Isometric Projection, each of the objects in Panels Nos. 15 and 16, on this page and Panels 1 to 6 on page 171.

All the drawings are in First Angle Projection and the dimensions are in millimetres.

They are not to scale.

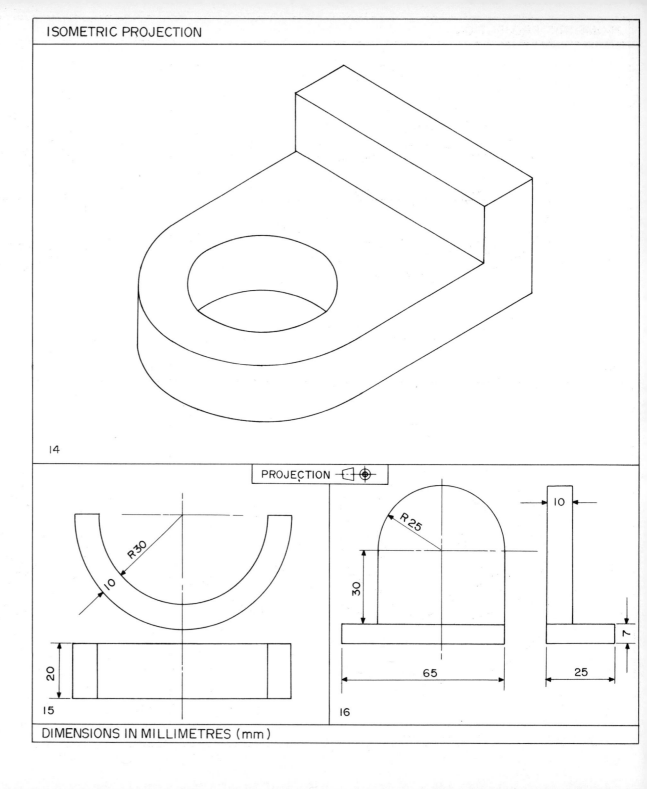

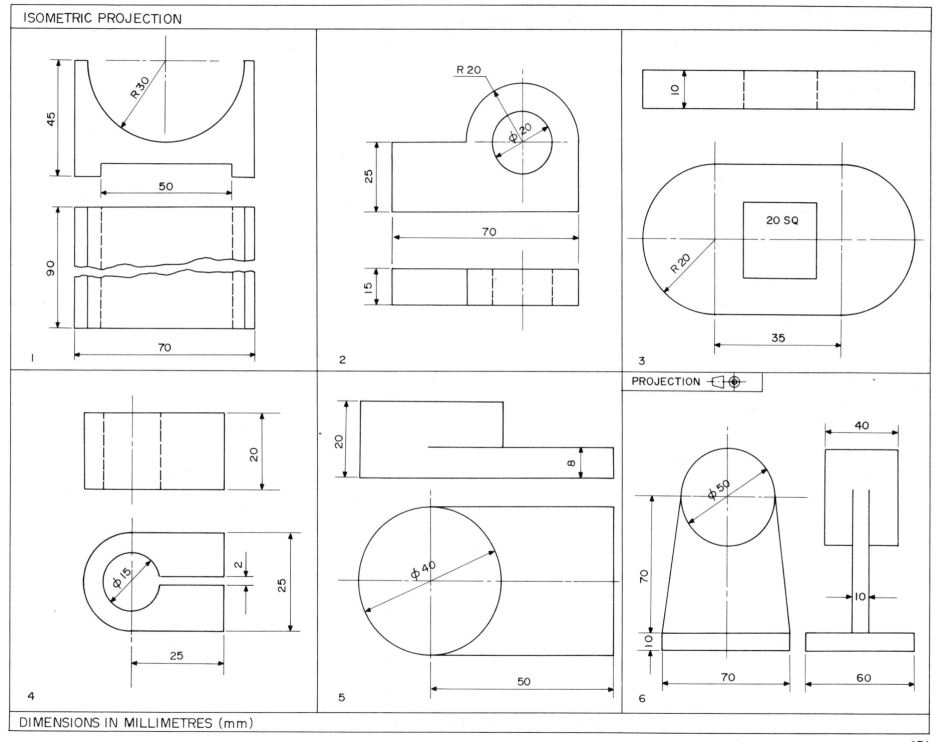

All the examples of curves in isometric projection have so far been approximations. This method produces a true ellipse but it has to be drawn freehand.

Draw the plate shown in Panel No. 1 following the instructions and drawings in the panels.

Panel No:

2   At the intersection of centre lines EF and GH draw a circle of 40 mm diameter. At 5 mm intervals draw the vertical ordinates X-Y-X^1 etc.

3   Draw the enclosing prism of the plate and centre lines EF and GH. Set out the 'square' ABCD with sides equal to the diameter of the hole. Mark E, F, G and H which are points on the ellipse.

4   5 mm Either side of the centre line EF draw vertical ordinates corresponding to those in Panel No. 2. With compasses or dividers set to YX in Panel No. 2 fix points X and X^1 by marking from Y.

5,6 Repeat the process to obtain points V, V^1, T and T^1.

7   Through the points you have obtained draw freehand the required ellipse.

8,9 This shows the projection of points L and N to obtain the visible part of the ellipse on the back of the plate. Further points are obtained by repeating the process.

10 Join the points with a freehand curve. Note that more of the curve has been drawn than is required to avoid 'kinking'. Visible edges are now lined in.

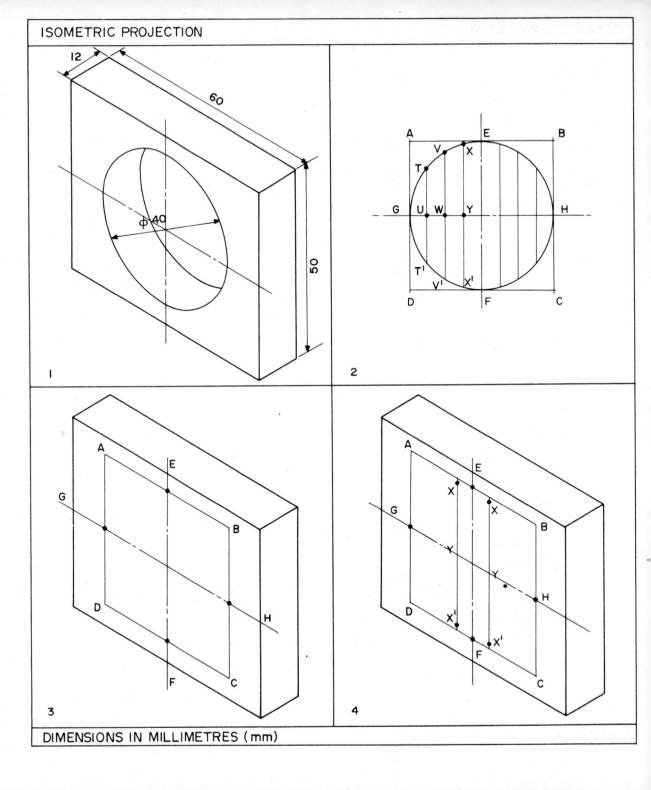

ISOMETRIC PROJECTION

DIMENSIONS IN MILLIMETRES (mm)

ISOMETRIC PROJECTION

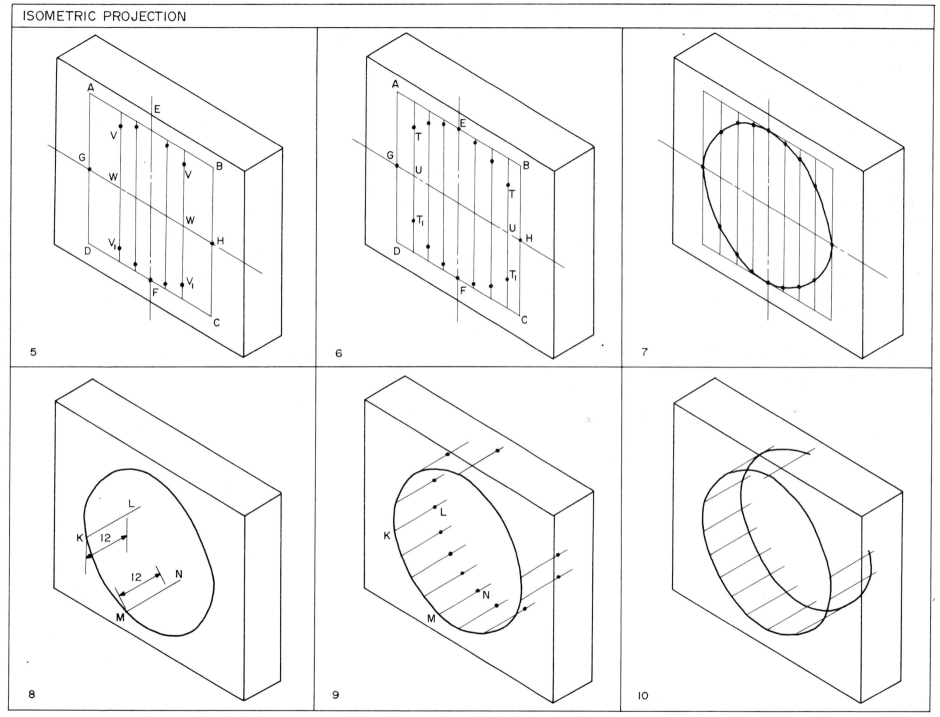

To draw the bracket shown in Panel No. 1.

Draw this exercise following, stage by stage, the instructions and the drawings in the panels.

Panel No:

1. Shows the plan and elevation of the bracket.

2. Shows a freehand pictorial drawing of the bracket.

3. Make an isometric drawing, full size, of the two containing prisms.

4. Set out, full size, a part plan of the curved part of the bracket.

   Insert centre lines EF and GH.

   At 10 mm intervals draw vertical ordinates XY and TU.

5. The drawings in this and the remaining panels are all to a larger scale.

   Draw the centre lines EF and GH and the ordinates at 10 mm intervals.

   E, F and G are points in the required curve.

6. Set your dividers or compasses to the distance YX on your part plan and mark points X on your isometric drawing. Repeat to obtain points T.

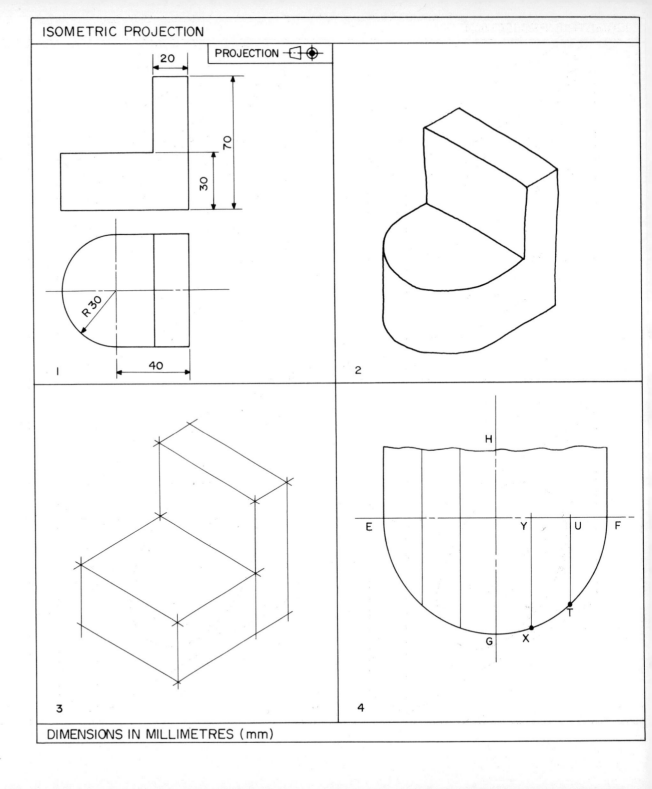

ISOMETRIC PROJECTION

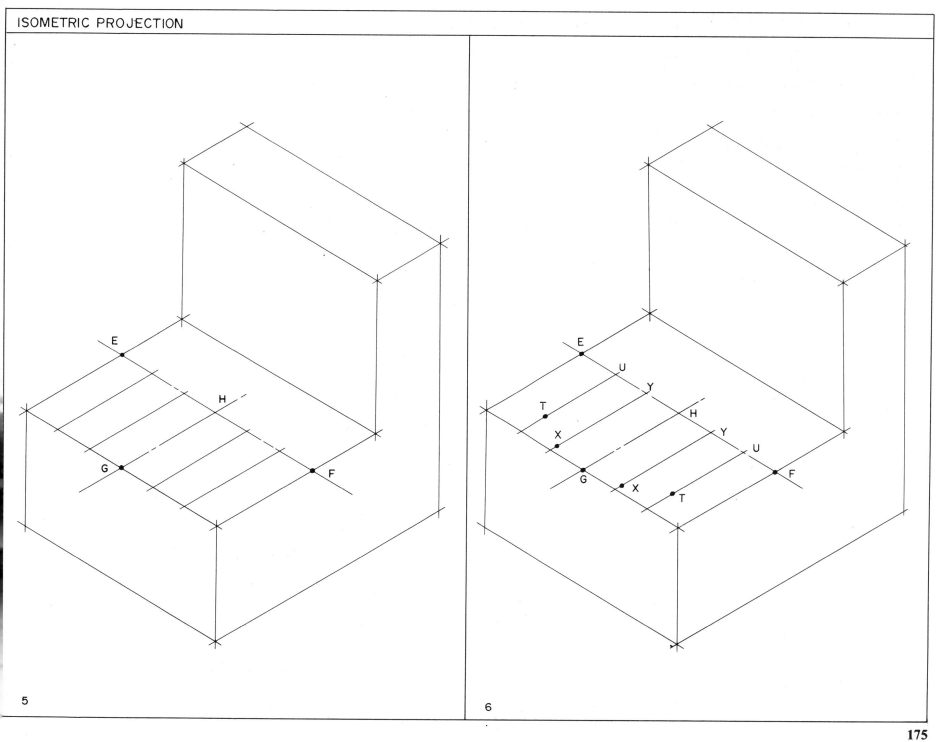

7. Through the points you have obtained draw a freehand curve.

8. To obtain the points for the bottom curve draw vertical lines downwards through F, S, T, X, G, V, W and U. Measure downwards along each of these lines 30 mm. (the thickness of the bracket) to fix points F_1, S_1, T_1, D_1, G_1, V_1, W_1 and U_1.

9. Join these points with a freehand curve. More of the curve than is required has been drawn to avoid 'kinking'. This would be required, of course, if hidden detail were to be shown.

10. Now line in your drawing.

    Notice that line LM is vertical and tangential to both curves.

Exercises:

Repeat this exercise varying the position of the bracket.

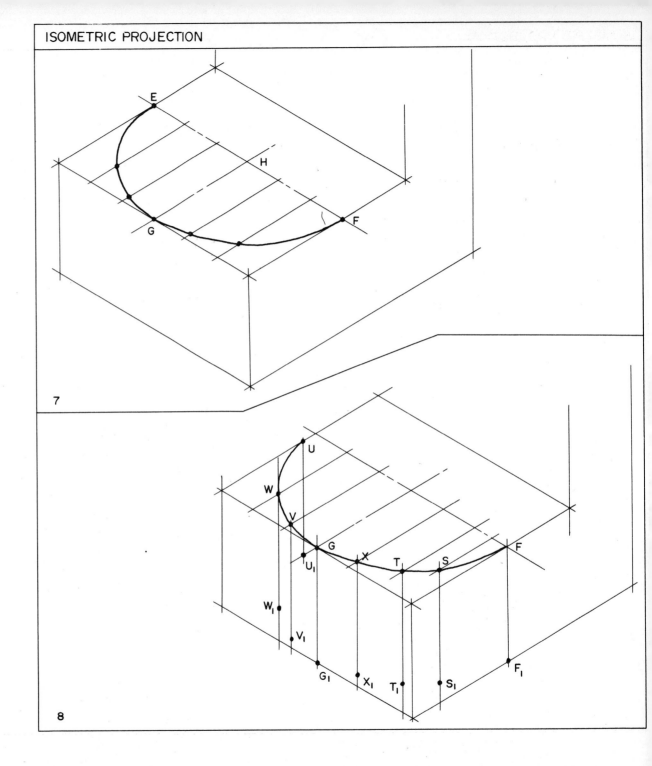

ISOMETRIC PROJECTION

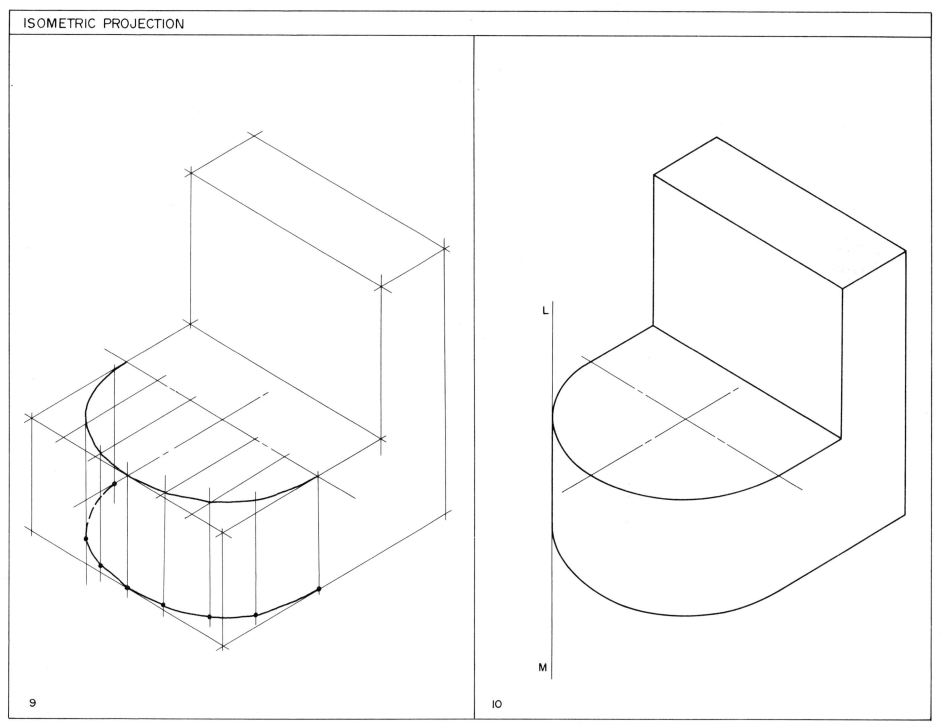

OBLIQUE PROJECTION

Oblique projection provides a pictorial view of an object showing three sides or faces. For small objects of uniform cross section it is somewhat easier to draw than isometric projection.

To draw the block shown in Panel No. 1 in oblique projection.

Draw this exercise following, stage by stage, the instructions and the drawings in the panels.

Panel No:

1. This shows the elevation and end elevation of a block.
2. Set out the base line and centre lines as shown to the dimensions given for the front elevation in Panel No. 1.
3. Complete the elevation.
4. Project the corners and the centres of circles at 45° as shown.
5. Set your compasses or dividers to 30 mm (the thickness of the block) and step off along the projectors and centre lines the thickness of the block.
6. With your compasses set at 20 mm and their point at 'O' draw in the part of the circle visible through the block.
7. With lines parallel to the front elevation draw in the back of the block as shown.
8. Complete and line in the drawing.
   (Note: In this exercise the full thickness of the block was drawn, however, it is usual to reduce measurements along the 45° line to half scale).

Exercise:

9. Make an oblique drawing of the vee block.

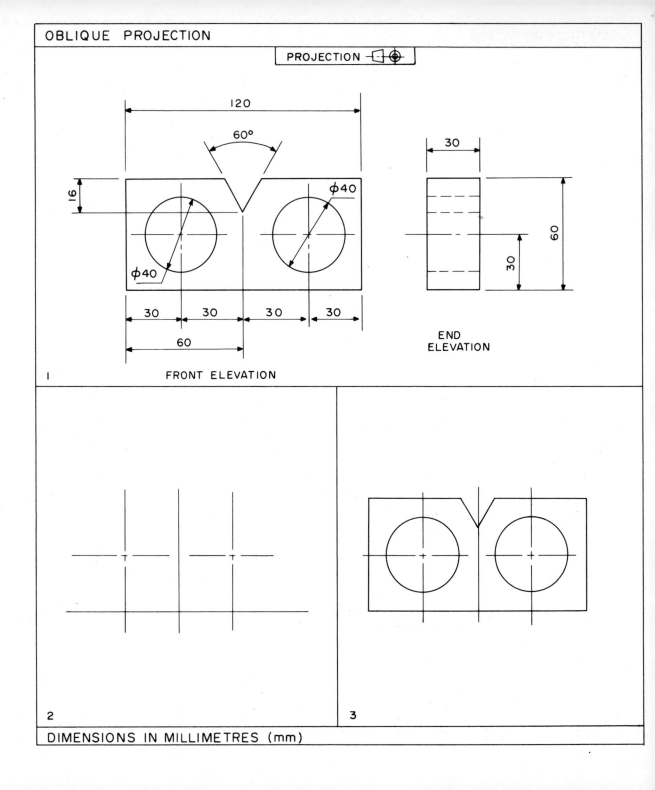

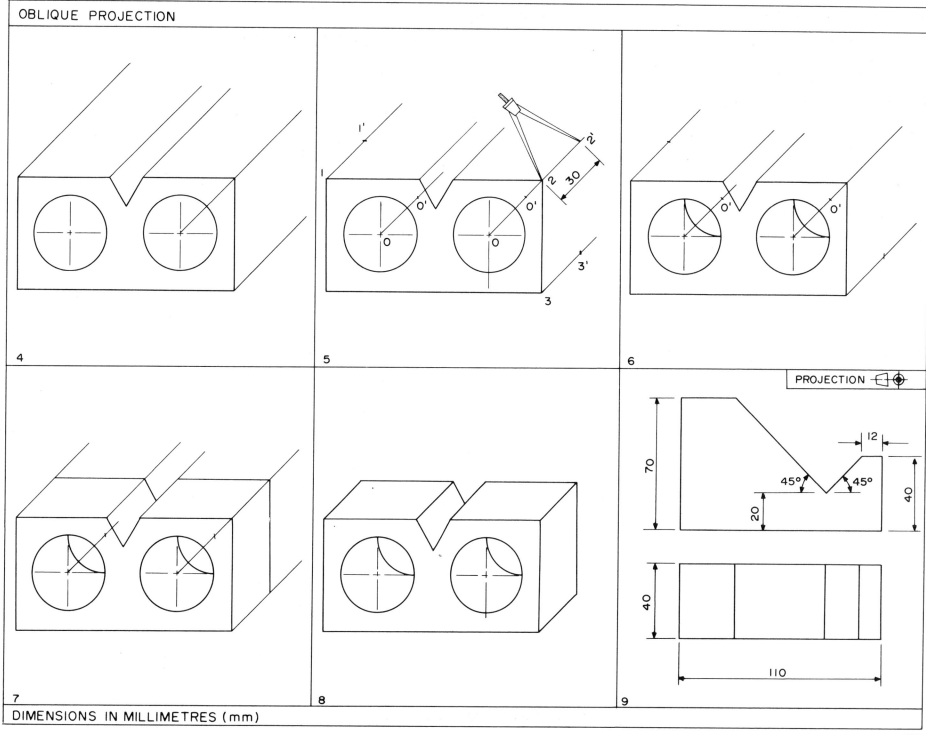

When drawing in oblique projection select the viewpoint which gives an end face containing the most detail (e.g. circles and curves, etc.).

To draw a cylinder in oblique projection.
Draw this exercise following, stage by stage, the instructions and the drawings in the panels.

Panel No:
1. The elevations and dimensions of a cylinder are given.
2. Draw the centre lines for the end face, view E.
3. Project the centre line at 45° and mark the length of the cylinder along it.
4. Draw the circular front face of the cylinder and the end face faintly.
5. At 45° draw tangents to the ends of the cylinder as shown.
6. Line in the visible part of the end of the cylinder.

Exercises:
7,8 Draw the two objects shown in oblique projection. In each case making the view E parallel to the plane of projection.

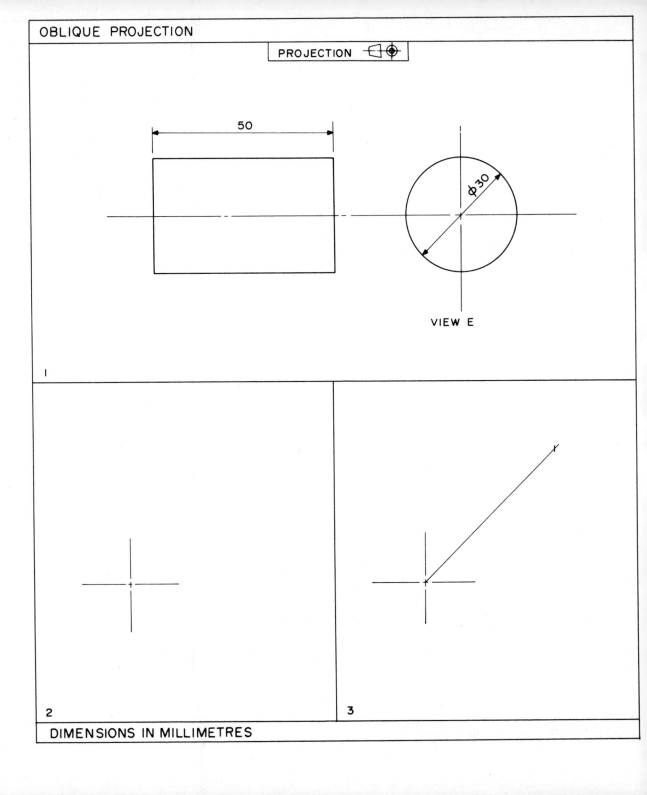

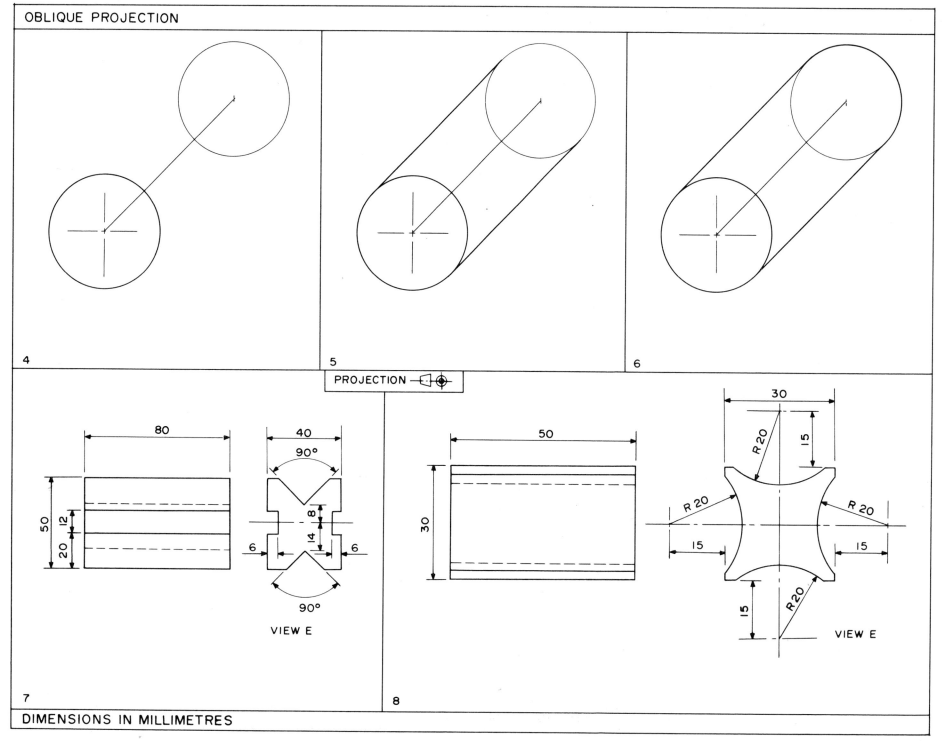

If an object with curves, circles or radii on its side or top has to be drawn in oblique projection then the curves are set out as for isometric projection.

To make an oblique drawing with a curve in the sloping side.

Draw this exercise following, stage by stage, the instructions and the drawings in the panels.

Panel No:

1 A dimensioned drawing of a bracket is given.

 The bracket is to be drawn in oblique projection with view E parallel to the plane of projection.

2 Draw the view E.

3 Project the corners at 45° and mark the length of the bracket as shown.

4 Draw the radiused corner as a separate construction dividing it into columns as indicated.

5 Transfer the widths 'w' to the oblique projection and draw the vertical lines.

6 Mark the heights h, h^1, h^2, h^3, etc.

7 With your tee square project each point and mark the thickness of the upright.

8 Join the points with freehand curves lining in the visible one.

9 Draw the tangent to both curves with your tee square and complete the lining in.

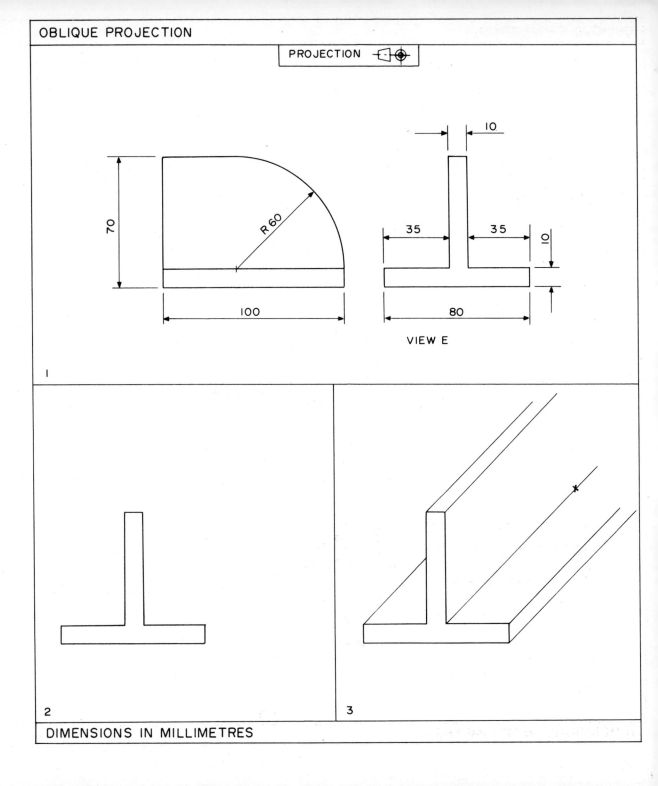

OBLIQUE PROJECTION

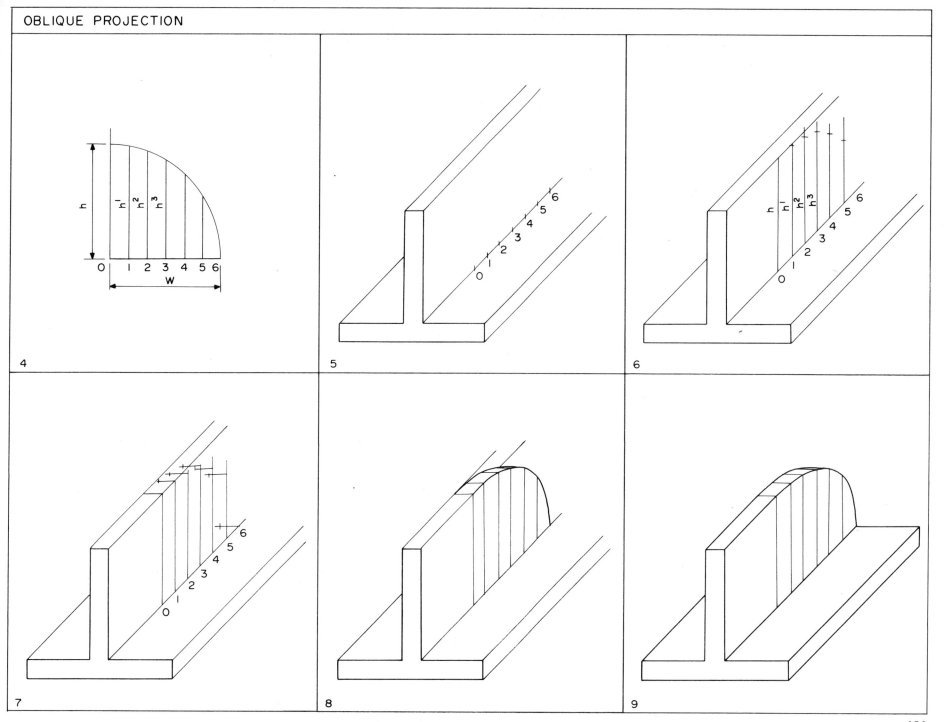

DEVELOPMENTS

The development of a solid consists of all its surfaces laid out flat. Every measurement on a development must be a true length. A correctly drawn development if cut out will fold up to enclose the original solid.

The development of the sides of a hexagonal prism is drawn as follows:

Panel No:

1. Draw the given plan and elevation of a hexagonal prism extending the ground line of the elevation and marking the width of the sides along it. Numbering the points to correspond with the plan reduces possibility of error.

2. For each point draw a perpendicular and project the height from the elevation. The complete development requires the addition of a hexagon for the top and base as indicated.

The development of the sloping sides of a square pyramid is constructed as follows:

3. Draw the plan and elevation of the given square pyramid. The true length of the sloping edge is found and re-drawn away from the elevation. With the true length as radius draw an arc.

   With the compasses set to the length of the base edge mark round the arc, numbering to correspond with the plan.

4. Join the points as shown. The complete development includes the square base added to one of the sides as indicated.

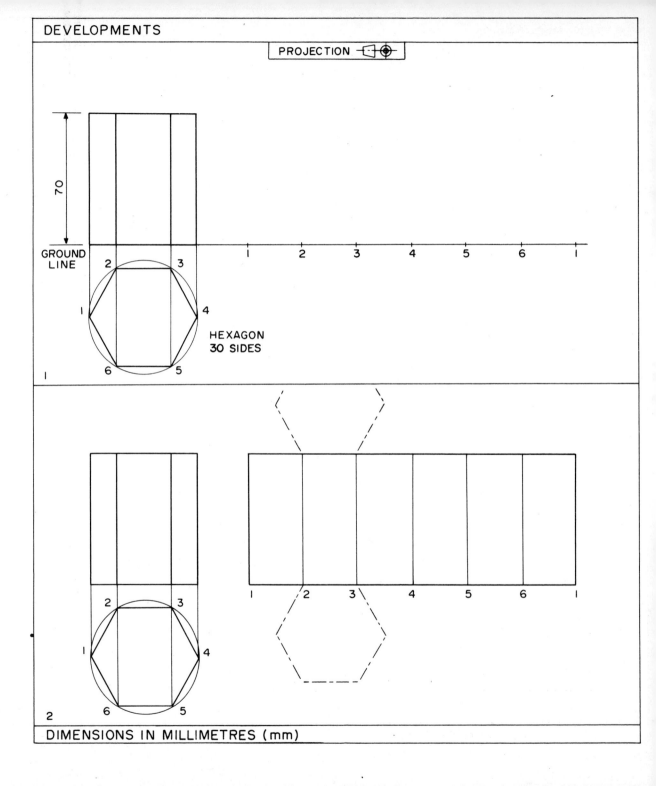

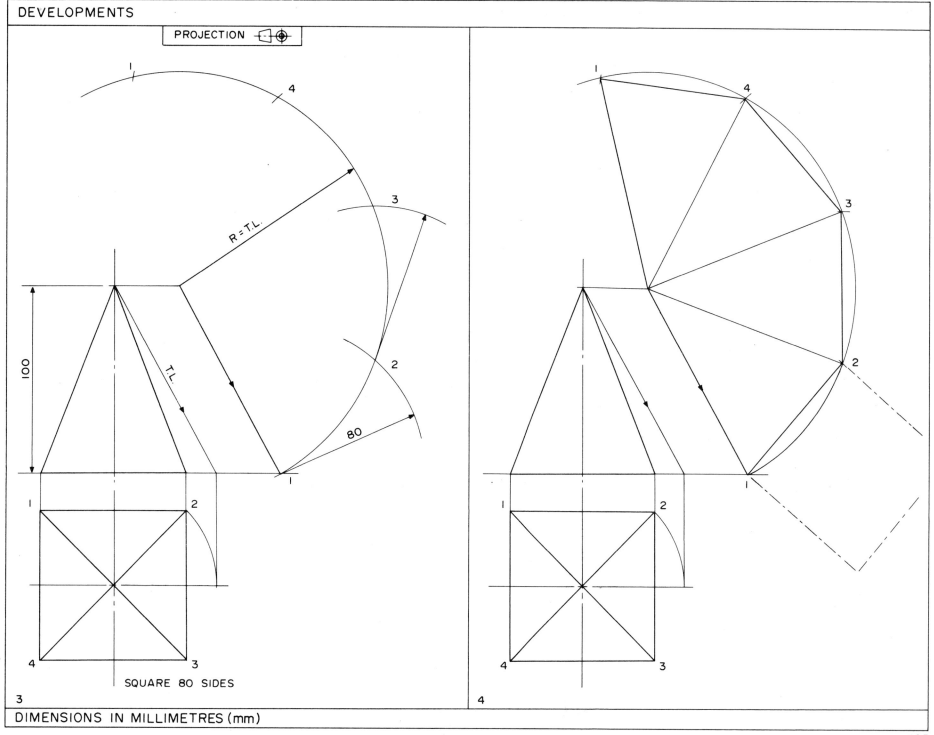

To draw the development of the sides of a container.
Draw this exercise following, stage by stage, the instructions and the drawings in the panels.

Panel No:
1. Draw the plan and elevation of the container to the dimensions given. On the extended ground line mark the width of the four sides and number them to correspond to the plan. Draw perpendiculars at each point.

2. The top of the container is projected into the development as shown. The lettered points in the elevation are above the numbered points on the base of the container and the same relationship can be seen in the development 'a' with 1, 'b' with 2, 'c' with 3, 'd' with 4.

To draw the development of the sloping side of a canopy or cover based on a square pyramid.
3. Draw the plan and elevation of the complete pyramid to the dimensions given. Find the true length of a slant edge and draw the development of the sides. Line in the lower part of the pyramid, complete the plan, lettering and numbering the corners as shown.

4. The remaining length of the slant edge is the same at each corner but it is the true length of a1, b2, c3 and d4 that is required. To obtain this project from 'd' in the elevation to intersect the true length.

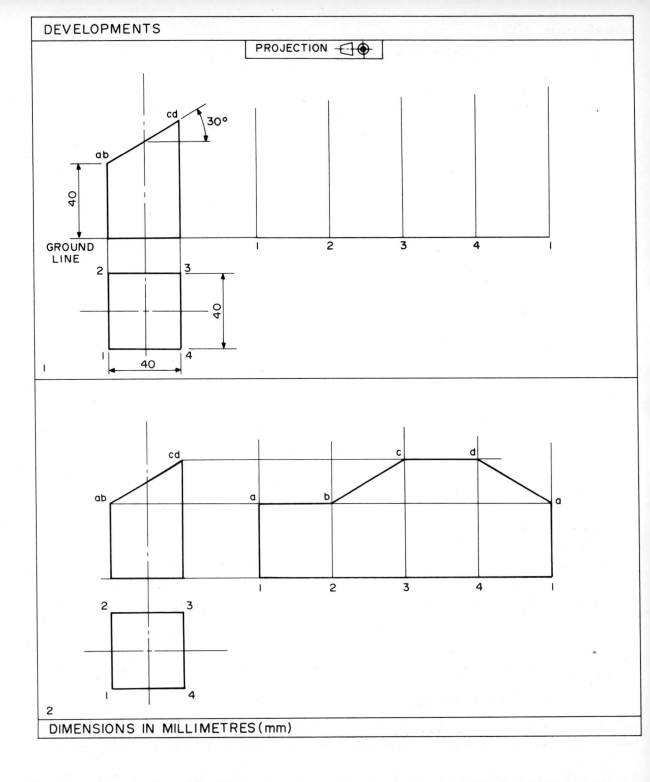

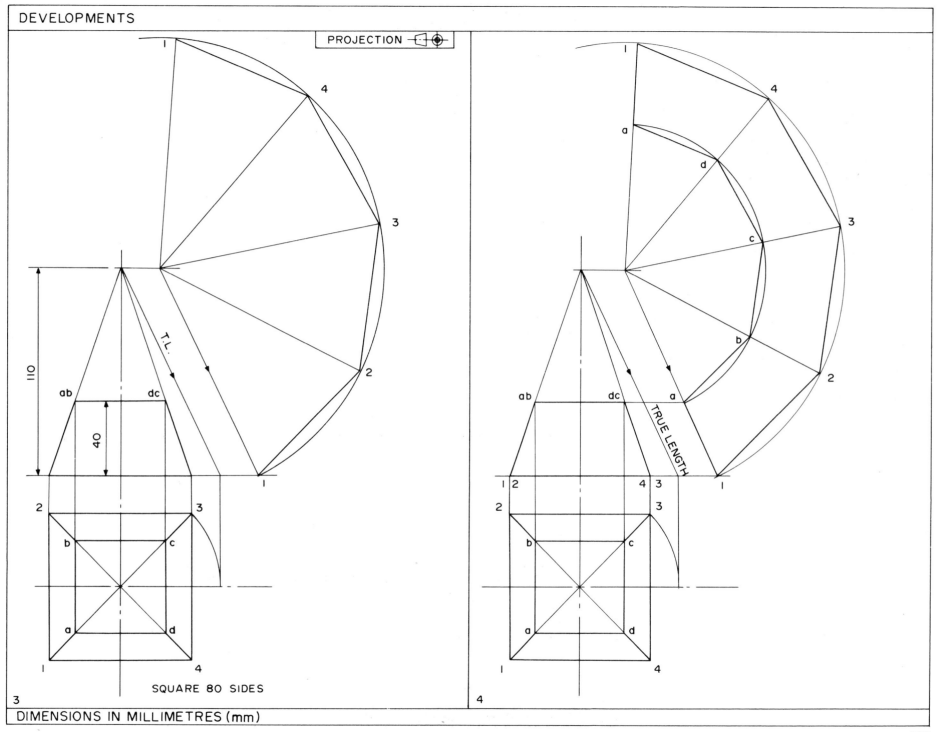

To develop the surface of a cylinder.
Draw this exercise following, stage by stage, the instructions and the drawings in the panels.

Panel No:
1. Draw the given plan and elevation of the cylinder. With your compasses divide the plan into twelve equal parts and number them as shown. Extend the ground along it by stepping out twelve equal units corresponding to 'w' a twelfth of the plan.

2. By projection from the elevation complete the development of the curved surface of the cylinder. The complete development includes the top and base of the cylinder as indicated.

To develop the surface of a cone.
3. Draw the given plan and elevation of the cone and divide the plan into twelve equal parts. Draw the slant height away from the elevation and with the slant height as radius draw an arc. With your compasses set to a twelfth of the circumference of the base mark round the arc as shown.

4. The development of the curved surface is finished as shown adding the circular base for the complete development.

These developments are only approximate but the method is sufficiently accurate for practical and drawing purposes.

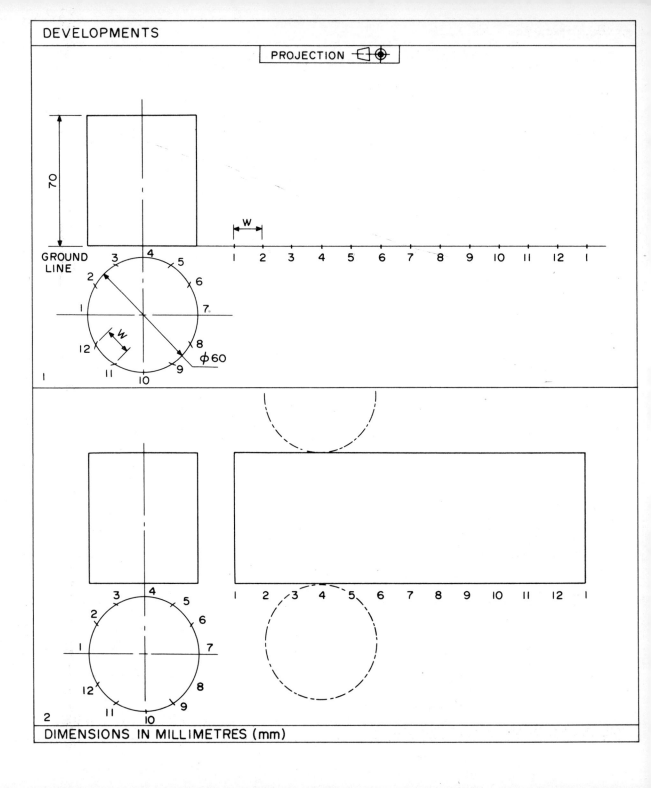

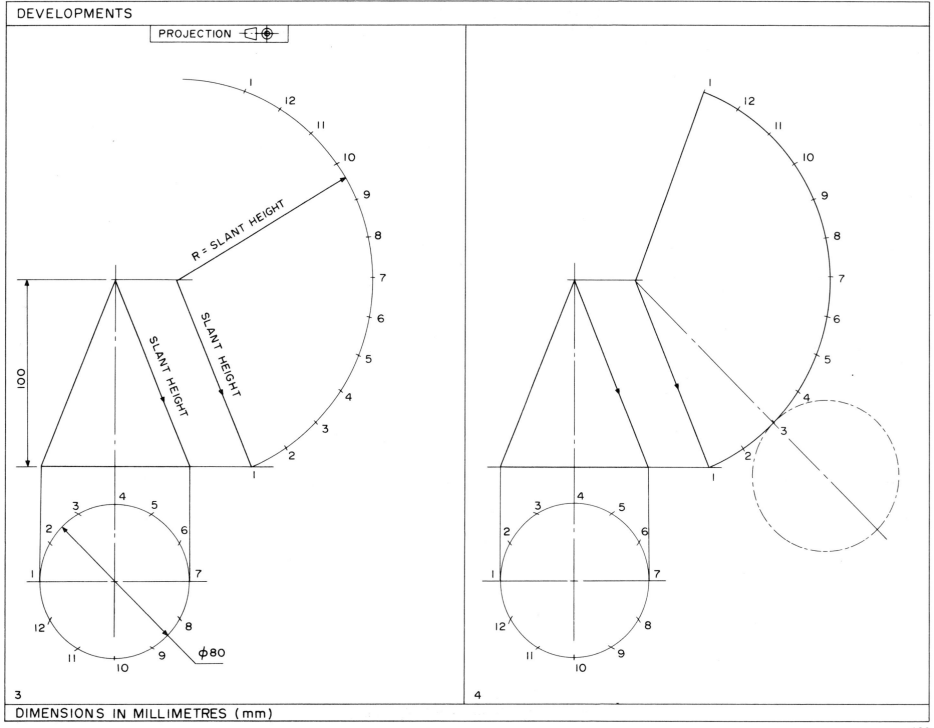

If a solid is cut by a plane it is said to be truncated. The remaining portion is called a frustum.

To draw the development of the curved surface of the frustum of a cylinder.

Draw this exercise following, stage by stage, the instructions and the drawings in the panels.

Panel No:

1. Draw the plan and elevation of the cylinder truncated by a plane at 45°. Divide the plan into twelve equal parts and project into the elevation numbering and lettering as shown. Mark the length of the circumference of the cylinder along the ground line and at each numbered point draw perpendicular lines.

2. Project the lettered points on the inclined surface onto the development. Each letter corresponds to a numbered point in the plan, a with 1, b with 2, c with 3, etc. The same relationship exists when the cylinder is unfolded and the lettered points are joined with a freehand curve.

To draw the development of the curved surface of the frustum of a cone.

3. Draw the plan and elevation of the cone truncated by a plane at 45°. Divide the plan into twelve equal parts and project into the elevation numbering and lettering as shown. Draw the development of the complete curved surface of the cone joining the numbered points to the centre of the arc. Every length on a development must be a true length so all the lettered points are projected onto the slant height.

4. The true lengths are now projected onto their corresponding numbered radii with a compass 'b' to 2, 'c' to 3, 'd' to 4, etc. Finally the points are joined with a smooth freehand curve.

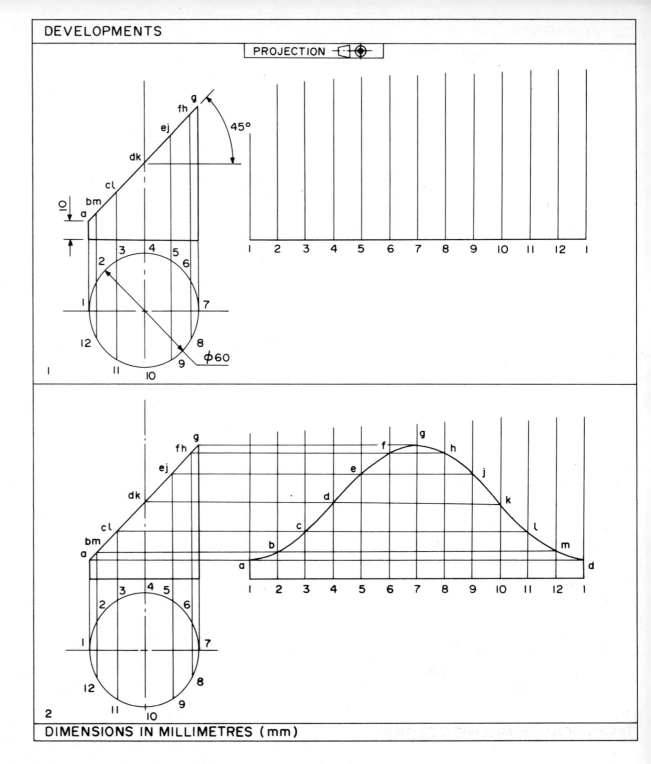

DEVELOPMENTS

DIMENSIONS IN MILLIMETRES (mm)

190

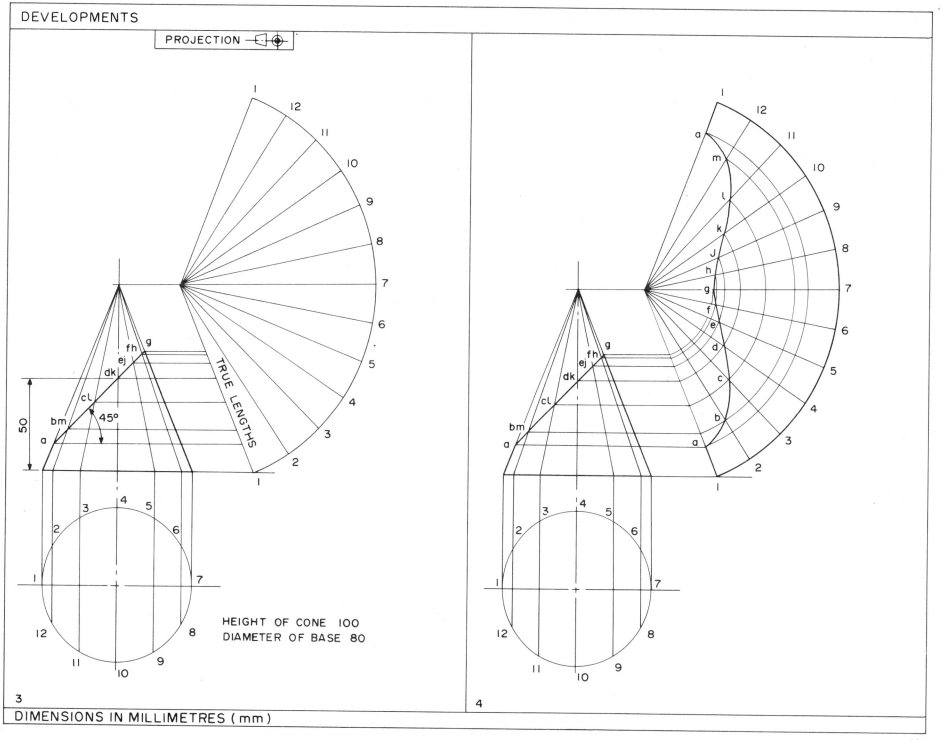

INTERPENETRATION

When two solids are joined as shown in Panel No. 1 the line of the joint is called the line of intersection.

Points on this curve are found by taking sections through both shapes as shown in Panel No. 2.

Sections are taken to produce shapes simple to draw e.g. rectangles, circles not ellipses parabolas, etc.

To draw the curve of intersection of two cylinders.
Draw this exercise following, stage by stage, the instructions and the drawings in the panels.

Panel No:
3 Draw the given views of the cylinders to the dimensions given.

4 Choose any horizontal section as indicated at AA fixing points 'a' and 'b'.

5 Project the points 'a' and 'b' into the plan as shown.

6 Points 'a' and 'b' are projected from the plan into the elevation — where only one is seen as the object is symmetrical.

7 Choose another horizontal section to fix points 'c' and 'd' which are projected into the plan and then into the front elevation as with points 'a' and 'b'.

8 If the horizontal sections are positioned at the same distance below the centre line as above it then, as can be seen, fewer projectors are needed. In this case 'a' and 'b' are the same distance above the centre line as 'e' and 'f' are below it.

9 Project sufficient points to obtain a good curve.

10 Draw in the complete curve of intersection.

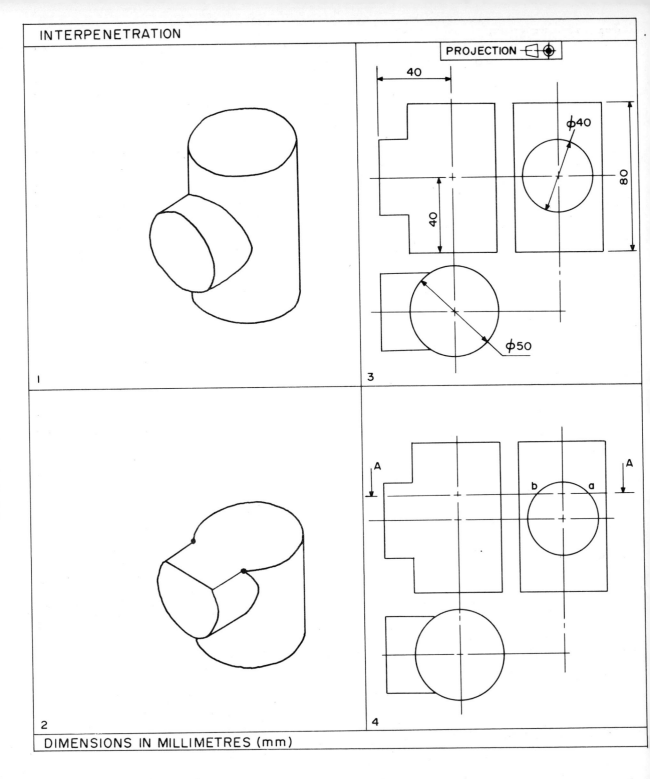

DIMENSIONS IN MILLIMETRES (mm)

INTERPENETRATION

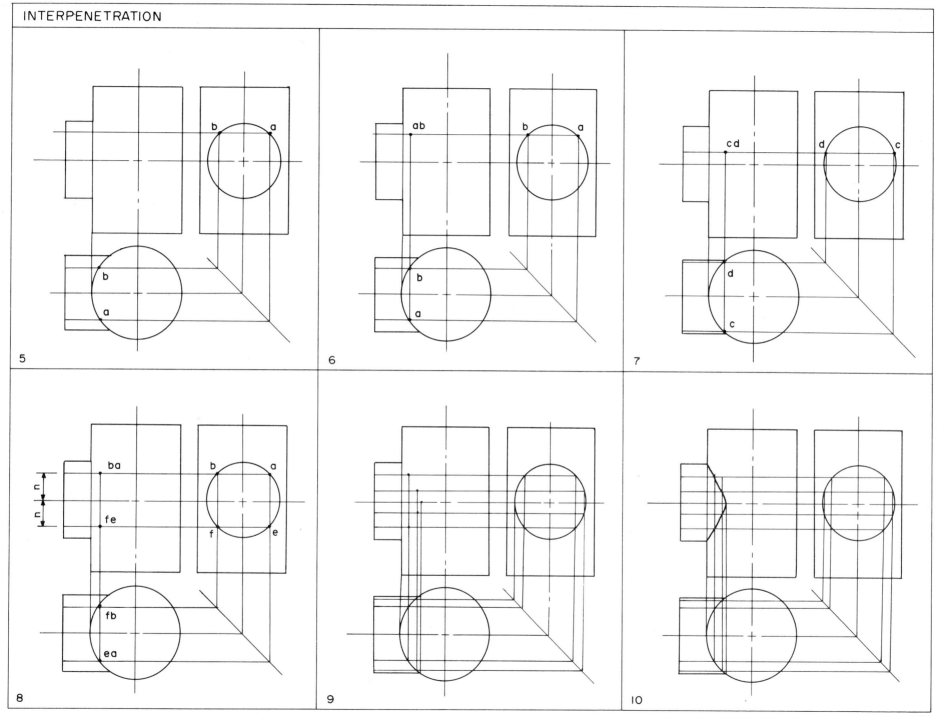

Time can be saved if half end views of the intersecting cylinders are drawn as in Panel No. 1.

Divide each end view into six equal parts either using a 60°/30° set square or by stepping off the radius as shown in Panel No. 2.

Panel No. 3 shows how to draw the curve of intersection.

To draw the curve of intersection of two cylinders with their axes offset.
Draw this exercise following, stage by stage, the instructions and the drawings in the panels.

Panel No:
4 Draw the given views using half elevations as above. Note the axes are offset.

5 'O' at the top of the horizontal cylinder is the mid point in plan.

Sections '2' are not only the same distance above and below the centre line in elevation but their width in plan is equal to 2–2 in the plan.

6 Continue plotting the points as shown but notice that there are two curves in elevation one of which is shown as hidden detail.

Exercise:
7-10 Draw lines of intersection for each of these examples.

What do you notice about No. 10?

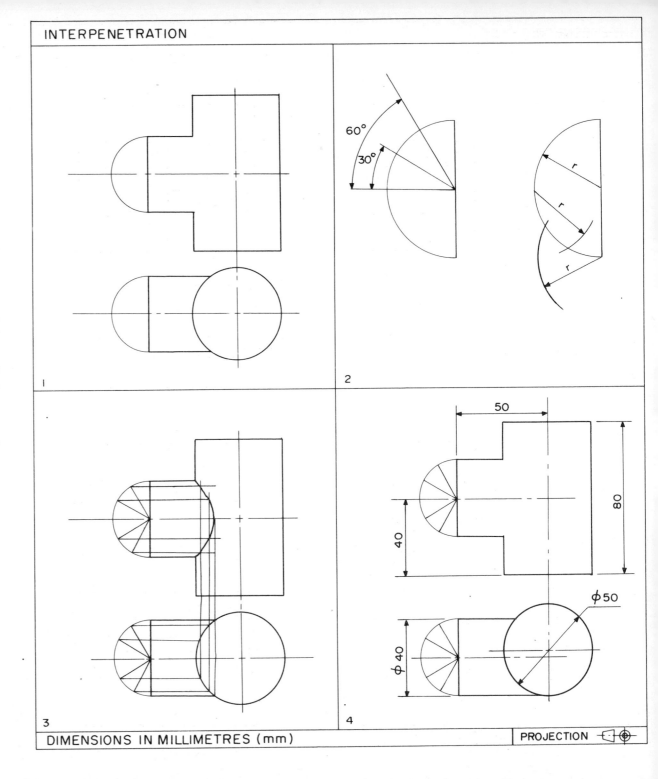

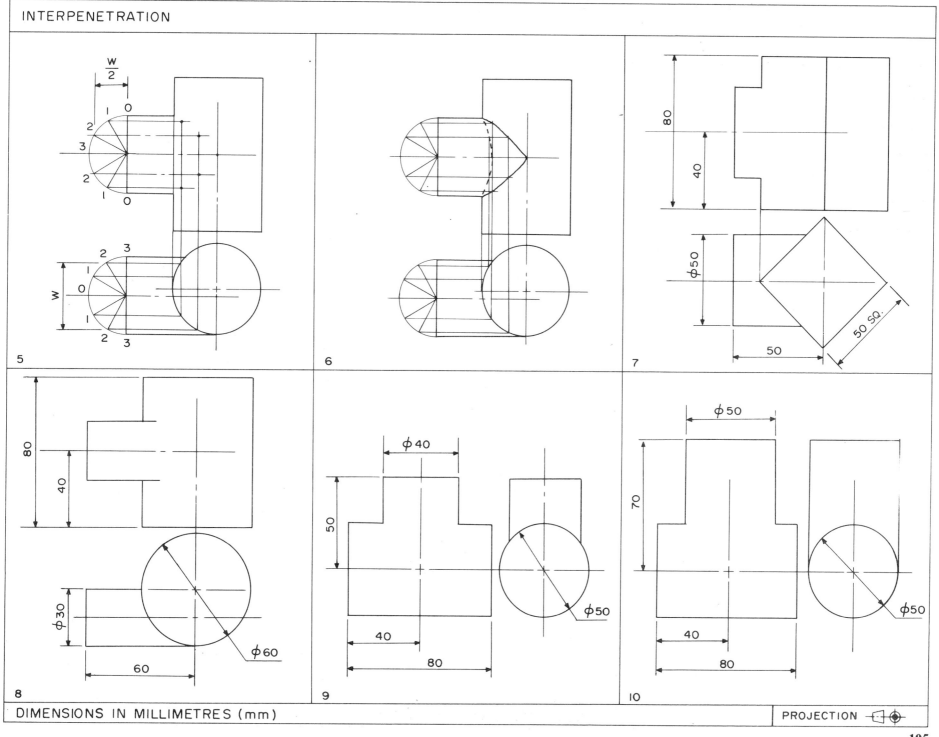

The line of intersection between a cylinder and cone can be found by taking a series of sections through both solids. The direction of the section is determined by the ease of drawing. In this example horizontal sections give a circle for the cone and rectangle for the cylinder in the plan, the intersection of these shapes provide the required points.

To draw the curve of intersection between a cylinder and a cone.
Draw this exercise following, stage by stage, the instructions and the drawings in the panels.

Panel No:
1. Draw the three views of the cylinder and cone to the given dimensions.
2. Draw a section line AA giving points 'a', 'b' and 'c'.
3. The point C where the section crosses the slant edges of the cone in elevation is projected into the plan giving C^1.

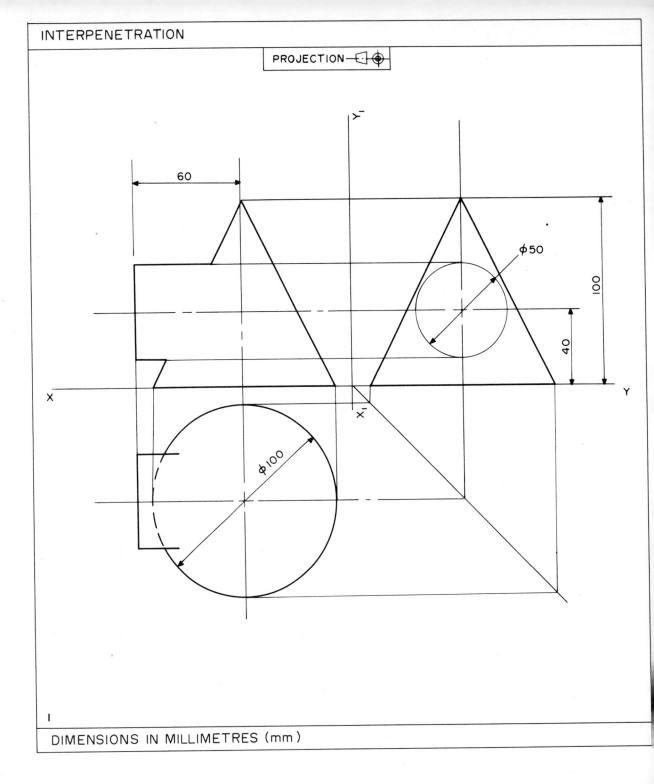

INTERPENETRATION

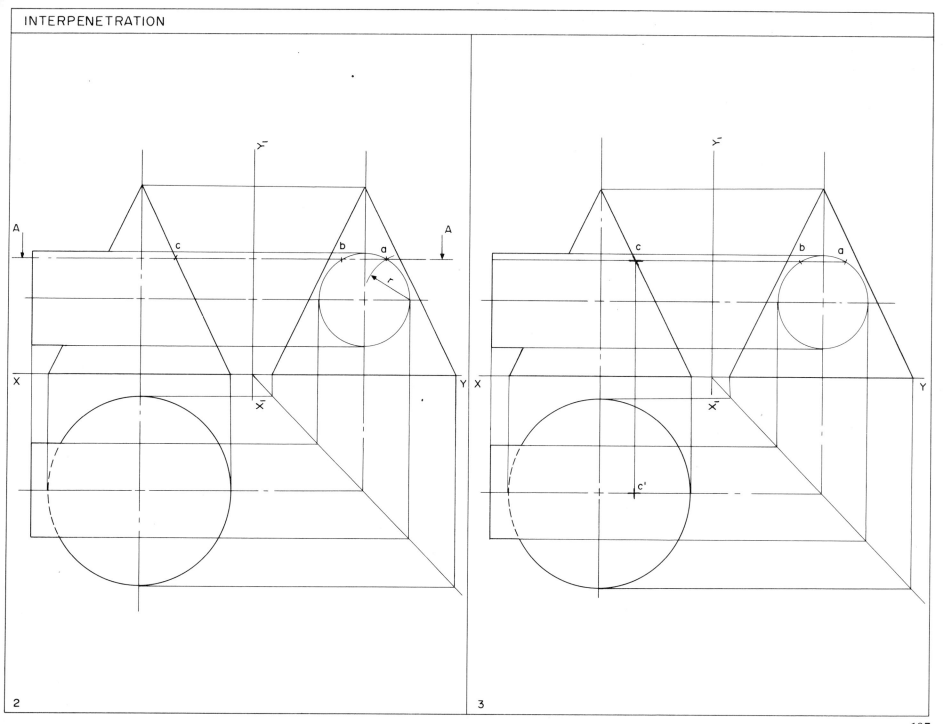

4 From the centre of the cone draw the circular cross section radius O-C^1 in the plan as shown.

5 The cutting plane in the end elevation gives the dimension 'w' the width of the section through the cylinder. The width 'w' is projected onto the change of direction line giving points a^1 and b^1.

6 The width 'w' is continued into the plan to intersect the circle giving points a^2 and b^2 on the curve of intersection.

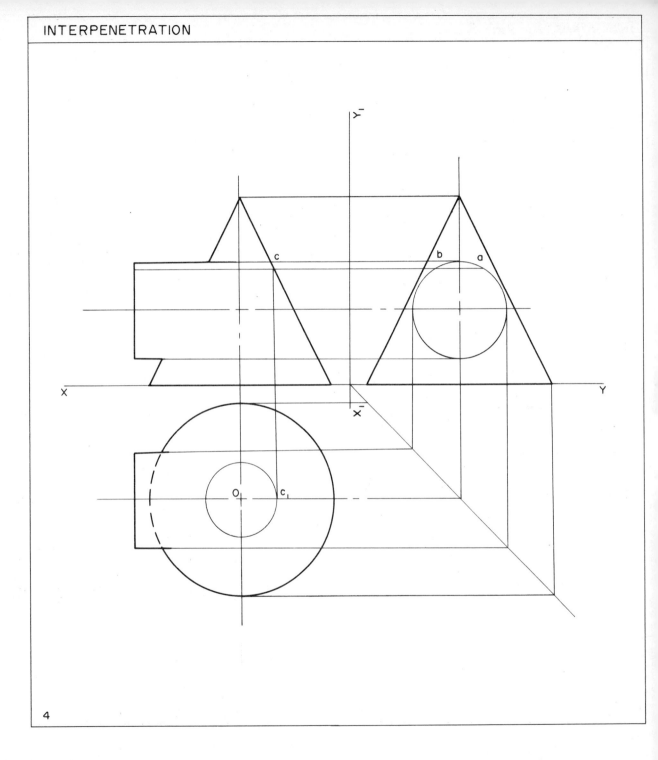

INTERPENETRATION

INTERPENETRATION

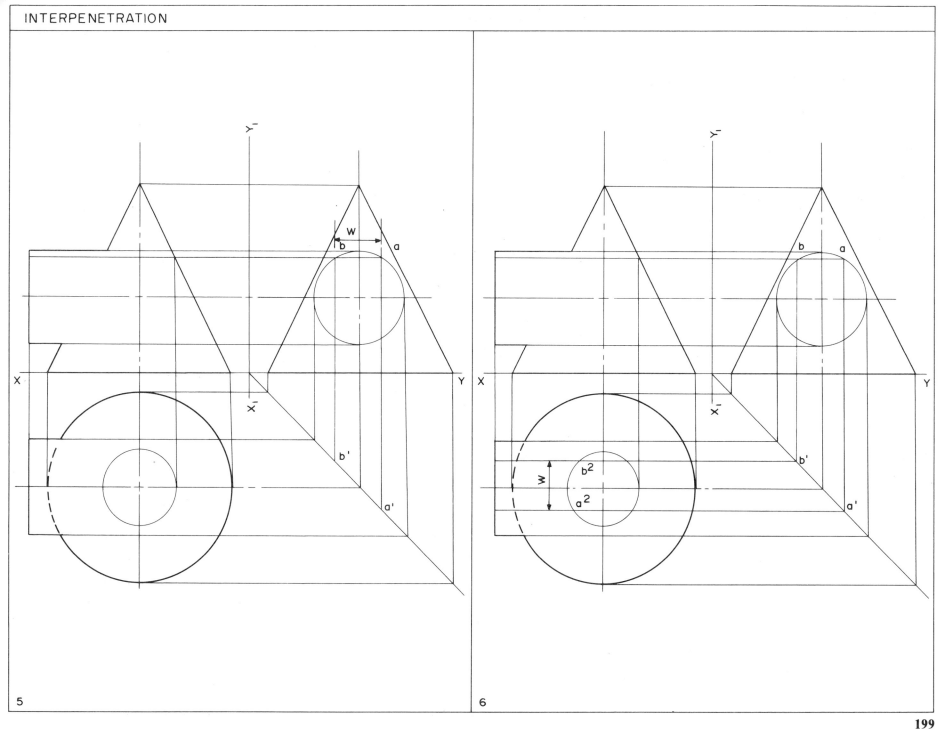

7 The points a^2 and b^2 are projected into the elevation and marked on the cutting plane in the elevation as shown.

8 To reduce the number of projected lines the next cutting plane BB is marked the same distance below the centre line as AA is above it. It fixes point 'd' and 'e'.

9 The intersection of the cutting plane with slant edge in the elevation fixing point 'f' is projected into the plan, and the circular cross section drawn radius $O\text{-}f^1$.

Its intersection with the cross section of the cylinder gives points d^2 and e^2.

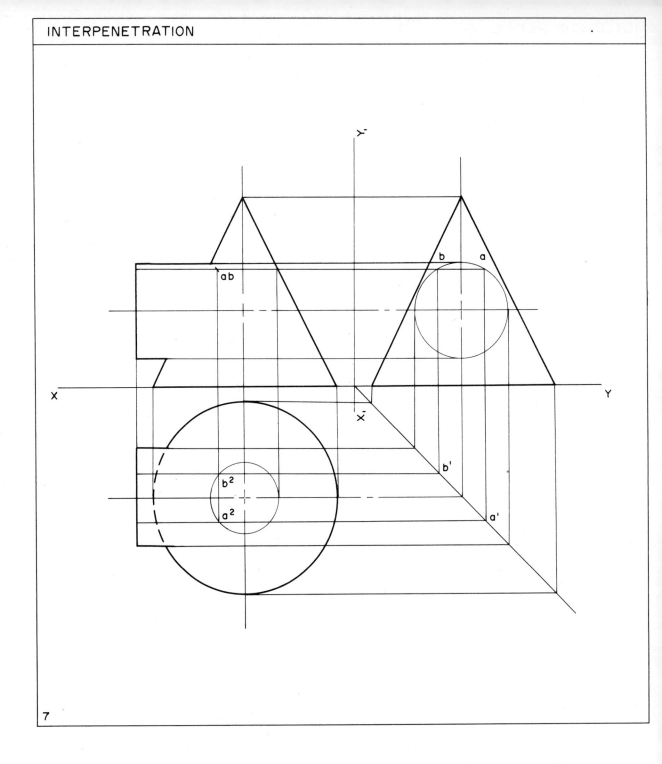

INTERPENETRATION

INTERPENETRATION

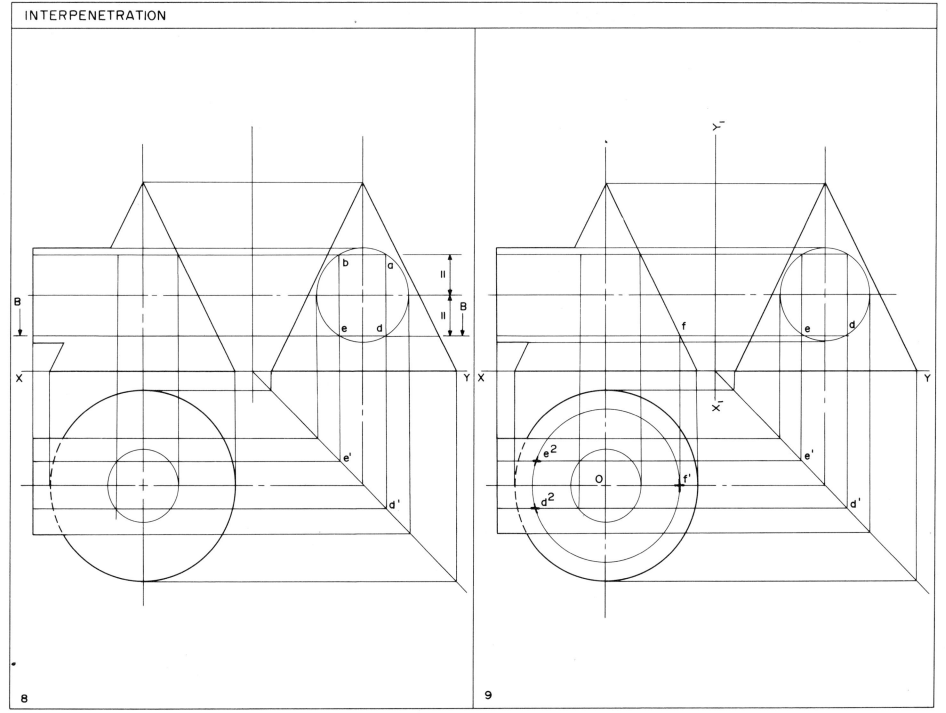

201

10 The points d^2 and e^2 are projected into the elevation. The ends of the curve of intersection in the elevation are projected into the plan to give 't' and 'u'.

11 Two more sections are fully projected as shown to give sufficient points to draw a fair curve in elevation and plan.

12 The points are joined as shown. The half end views divided into six equal parts and numbered as shown can be used to draw the curves of intersection without a full end elevation.

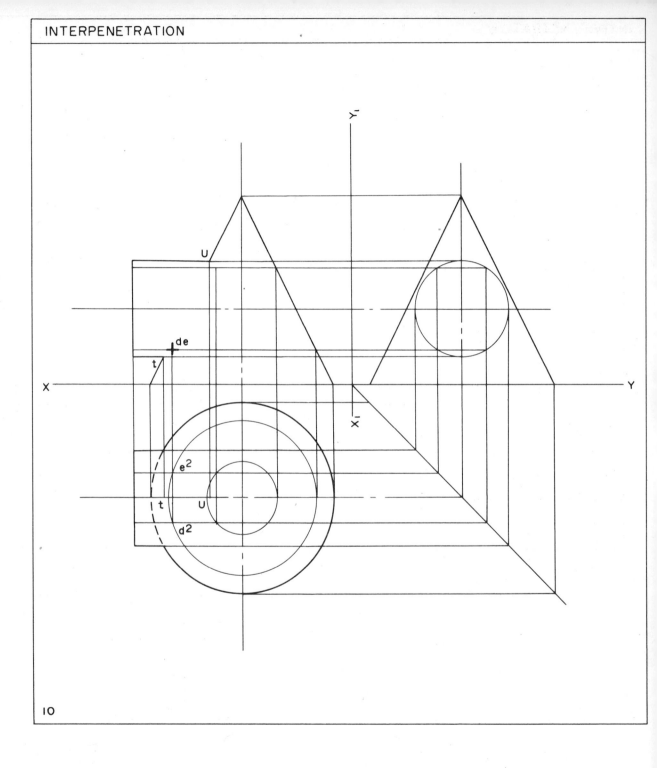

INTERPENETRATION

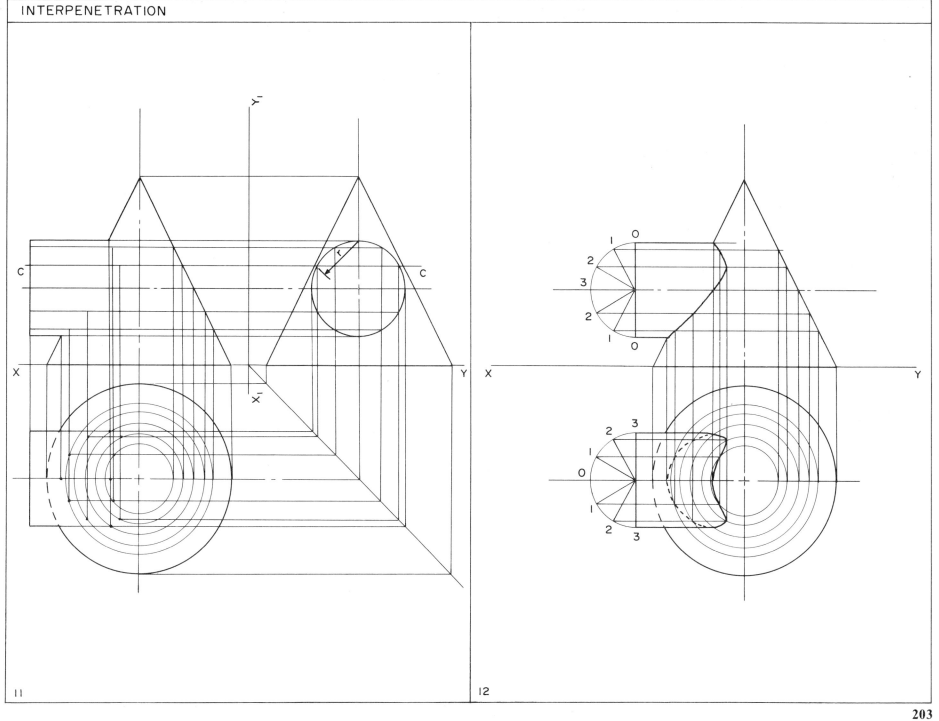

TRUE LENGTH

A line which is inclined to both the horizontal and the vertical planes is said to be oblique. Neither its elevation nor its plan will show its true length or the true angles it makes with these planes. These can be found by drawing.

To find the true length of an oblique line by auxiliary projection.

Draw this exercise following, stage by stage, the instructions and the drawings in the panels.

Panel No:

1. Draw the plan and elevation of the given line.
2. Draw a second XY line parallel to the plan of line 'ab'. Draw also a change of direction line CD.
3. Project points 'a' and 'b' from the plan at right angles to the new XY line.
4. Using the change of direction line CD project the heights of 'a' and 'b' from the front elevation into the new elevation.
5. Join 'a' and 'b' in the auxiliary projection as shown. This gives the true length of 'ab' and its true angle θ with the horizontal plane.

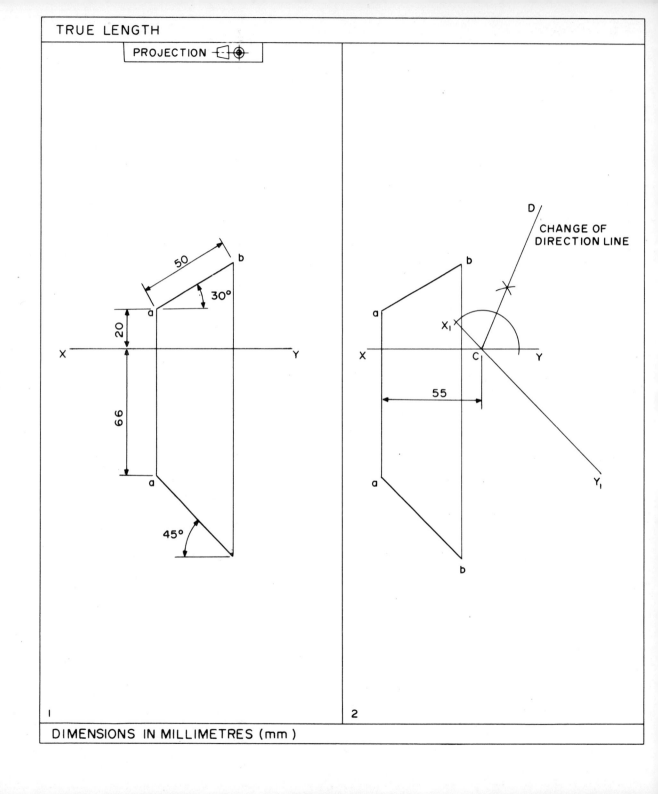

TRUE LENGTH

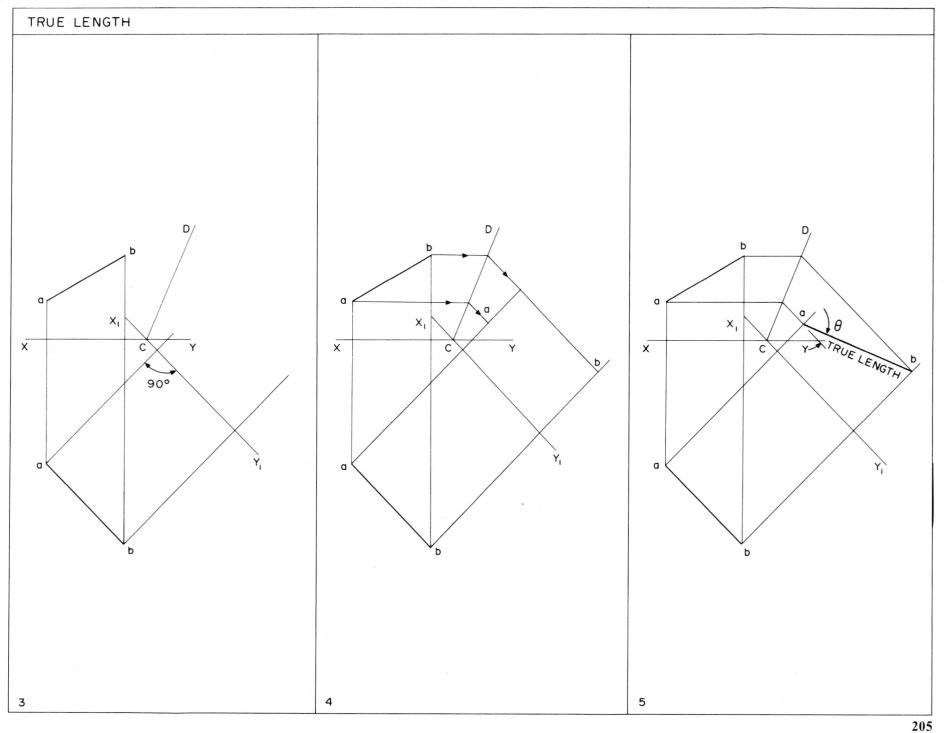

To find the true length of an oblique line by auxiliary projection.

Draw this exercise following, stage by stage, the instructions and the drawings in the panels.

Panel No:

1. The true length of the line ab and its inclination to the vertical plane can be found by drawing an auxiliary plan on a plane parallel to the elevation of ab. Draw the plan and elevation of the line ab to the given dimensions including the XY line in the position shown.

2. Draw a second XY line parallel to the elevation of 'a', 'b', as shown and include the change of direction line CD.

3. Project from the elevation at right angles to X^1Y^1 the points 'a' and 'b'.

4. Using the change of direction line project the points 'a' and 'b' from the plan into the auxiliary view.

5. Join the intersection of the projectors 'a' and 'b' to complete the auxiliary plan giving the true length of the line ab and its inclination to the vertical plane angle ∅.

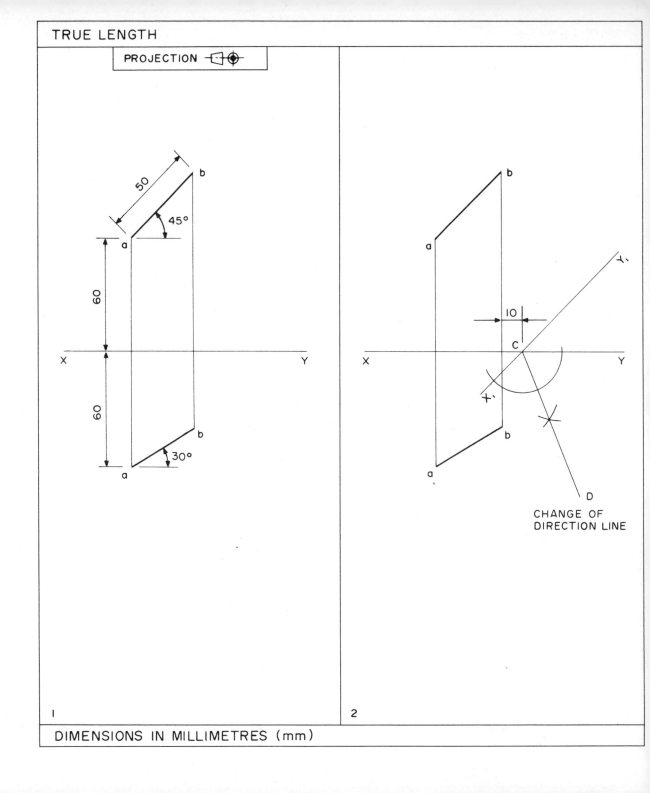

DIMENSIONS IN MILLIMETRES (mm)

TRUE LENGTH

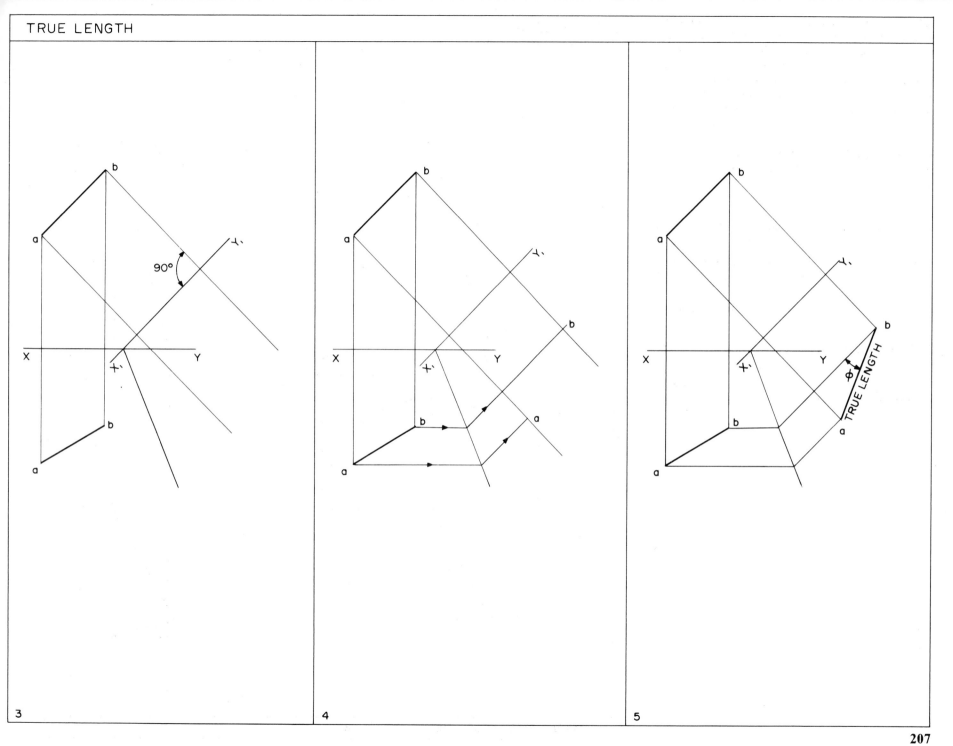

3

4

5

207

An alternative to finding the true length of a line by auxiliary projection is to consider the line as a generator of a cone. The plan and elevation of a cone is shown in Panel No. 1 with a generator 'ab' lined in. The true length of the line since it is on the surface of the cone is the length of the slant edge. The slope of the slant edge is the true angle with the horizontal plane.

To find the true length of an oblique line.
Draw this exercise following, stage by stage, the instructions and the drawings in the panels.

Panel No:
2 Draw the plan and elevation of ab to the given dimensions including the XY line.

3 With centre 'b' and radius 'ba' draw an arc to give 'ba_1', parallel to the XY line. (Part of the base of the imaginary cone).

4 In elevation 'a' is extended horizontally parallel to the XY line.

5 Project 'a' in plan into the elevation.

6 Join 'b' to 'a_2', to give the true length of 'ab' and θ its angle with the horizontal plane, (slant edge of the imaginary cone).

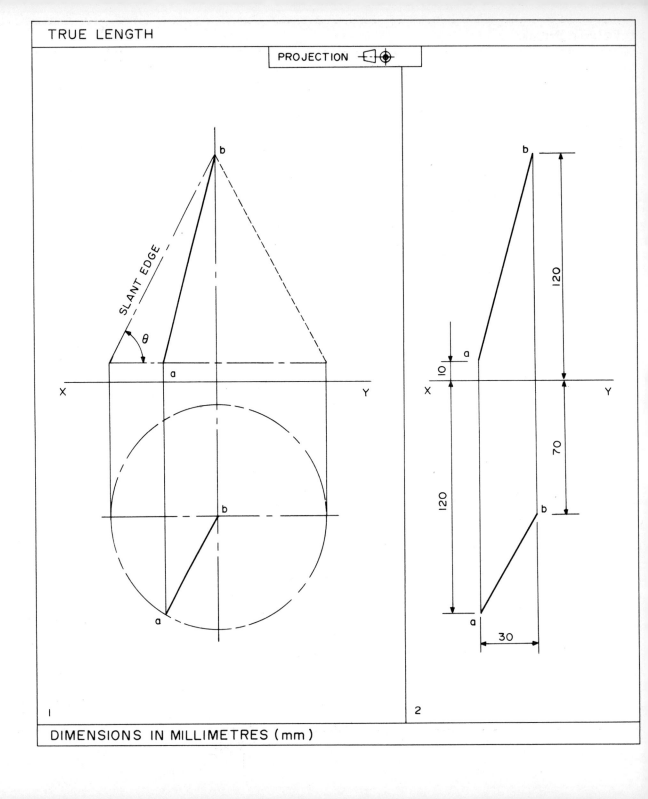

DIMENSIONS IN MILLIMETRES (mm)

TRUE LENGTH

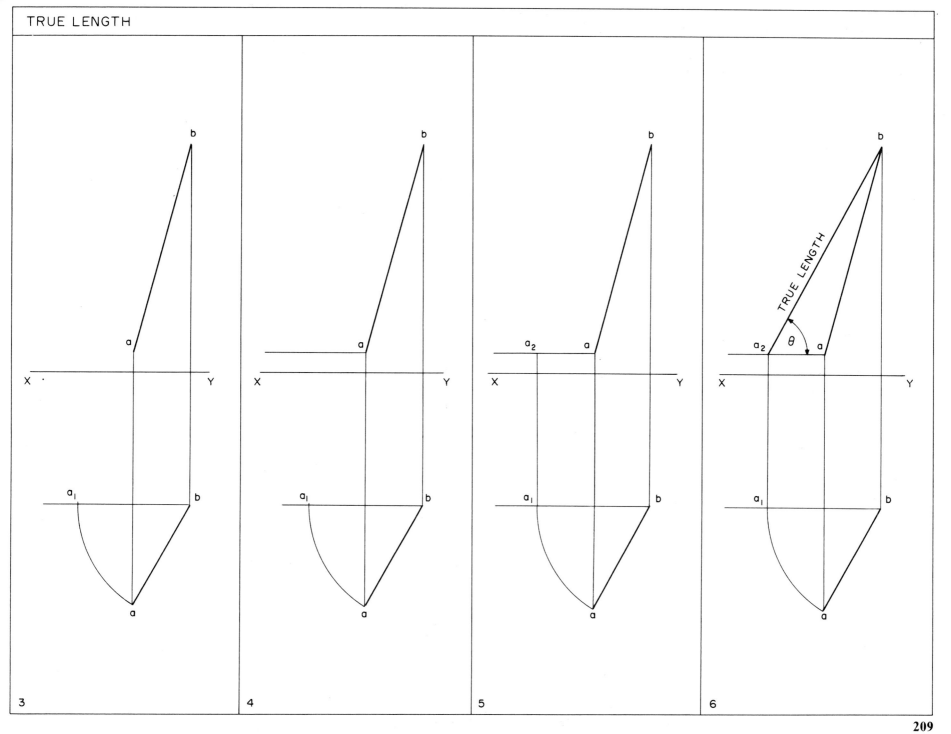

If the line AB is considered as a generator of a cone, as shown in Panel No. 1, which has its base parallel to the vertical plane then the angle ∅ is the true angle of the line to the vertical plane and the slant edge is its true length.

To find the true length of an oblique line and its true angle with the V.P.
Draw this exercise following, stage by stage, the instructions and the drawings in the panels.

Panel No:
2 Draw the plan and elevation of ab to the given dimensions including the XY line.

3 With centre 'b' and radius 'ba' draw an arc to give 'ba_1', parallel to the XY line (Part of the base of the imaginary cone).

4 In plan 'a' is extended parallel to the XY line.

5 Project 'a_1' in the elevation into the plan.

6 Join 'b' to 'a_1' to give the true length of 'ab' and ∅ its angle with the vertical plane.

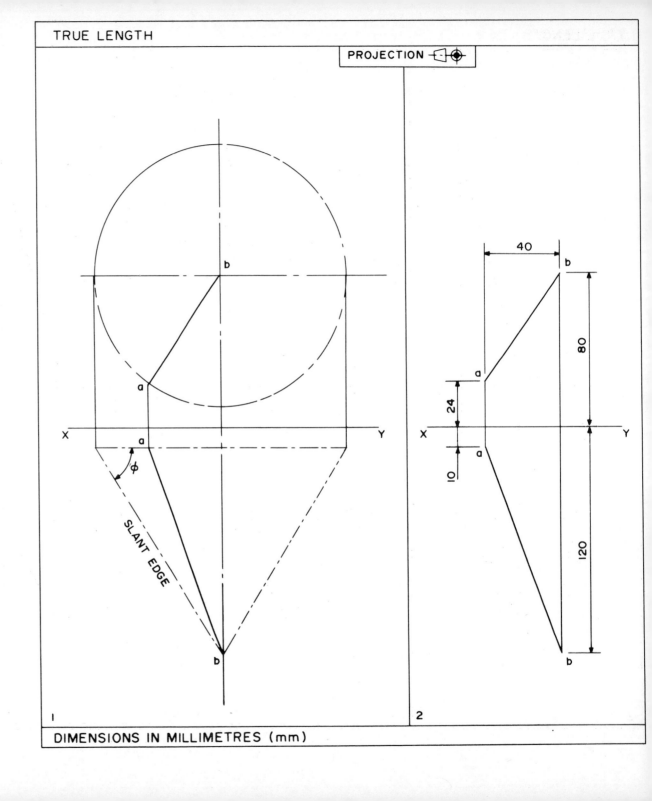

DIMENSIONS IN MILLIMETRES (mm)

TRUE LENGTH

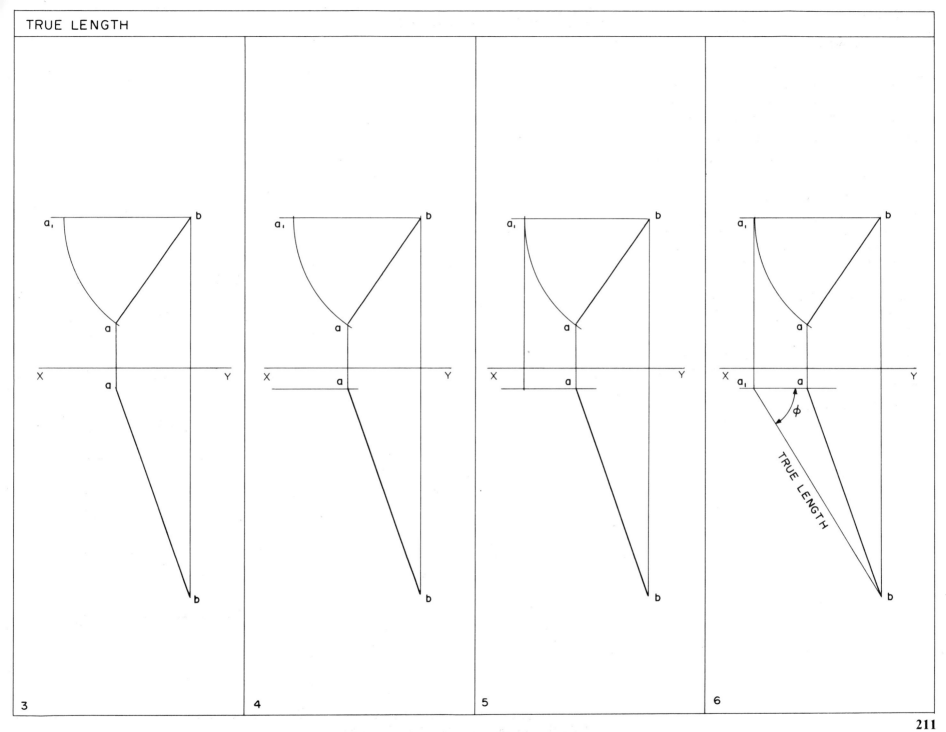

EXAMPLES

Construction of angles.

Panel No:

1. On the given line BC, construct an angle CBA of 120° and an angle BCD of 105°. Find the centre of a circle touching all three lines. Draw the circle.

   West Yorks & Lindsey Regional Ex. Board.

2. Draw a right angle triangle with AB as the hypotenuse.

   Southern Regional Ex. Board.

3. The panel shows in diagrammatic form a pair of dividers with legs of 100 mm long. Copy this drawing showing the construction you have used to obtain the angle of 75° (A protractor must *not* be used).

4. Draw the square ABCD as shown. By geometrical construction, find the mid point of each side of the square. Join up the four points with straight lines to produce a second square. Measure and write down the length of one of the sides of this square in millimetres correct to the nearest millimetre.

   North Western Sec. Schools Ex. Board.

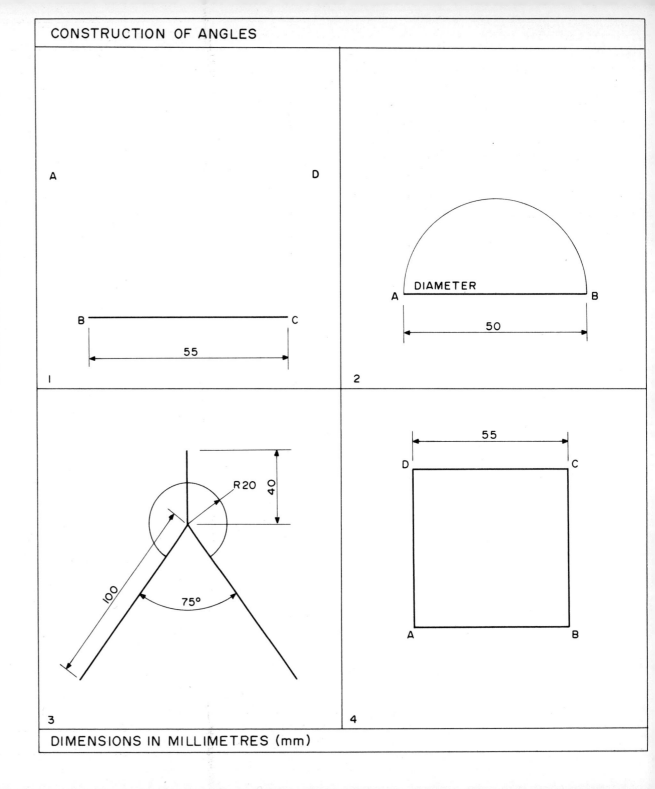

212

Construction of triangles

Panel No:

1. On a line AB construct an isosceles triangle having an area of 60 mm^2.

   N. Western Sec. Schools Ex. Board.

2. AB is one side of a triangle ABC. Angle ABC is 30° (construct with compasses) and side AC is 3/5 as long as AB. Show your construction to complete the triangle.

   E. Mid. Regional Ex. Board.

3. Line ABCD represents a piece of wire. The wire is to be bent to form a triangle with A and D coming together at the apex of the triangle. Draw the triangle.

   S. E. Regional Ex. Board.

4. BC is the hypotenuse of a right angled triangle side AC is 45 mm in length. Construct the triangle.

   West Yorkshire and Lindsey Regional Examining Board.

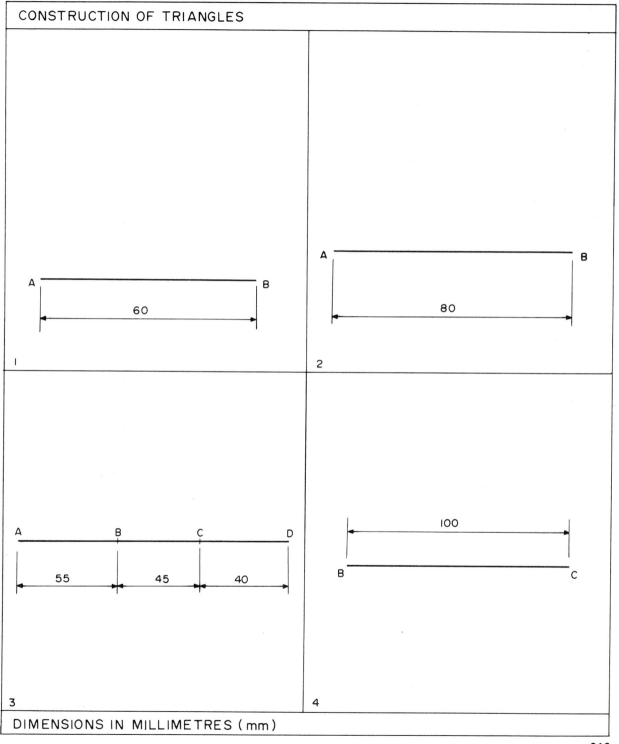

Division of lines

Panel No:

1. The panel shows a fish slice to be drawn to the given dimensions. The construction for finding the centre of the holes must be shown clearly.

2. Divide the given line AB in the proportions of 2 : 3.

    East Anglian Ex. Board.

3. Many of the numbers on a thermometer are no longer visible and you wish to make a cardboard scale which may be glued to the wooden backing. The figure represents the piece of cardboard onto which 2 marks which are visible have been transferred. Mark the scale in units of 10° from 0° to 110°. SHOW YOUR CONSTRUCTION.

    E. Midland Regional Ex. Board.

4. Divide line AB into three parts in the proportion 1 : 1½ : 2.

    N. Western Sec. Schools Ex. Board.

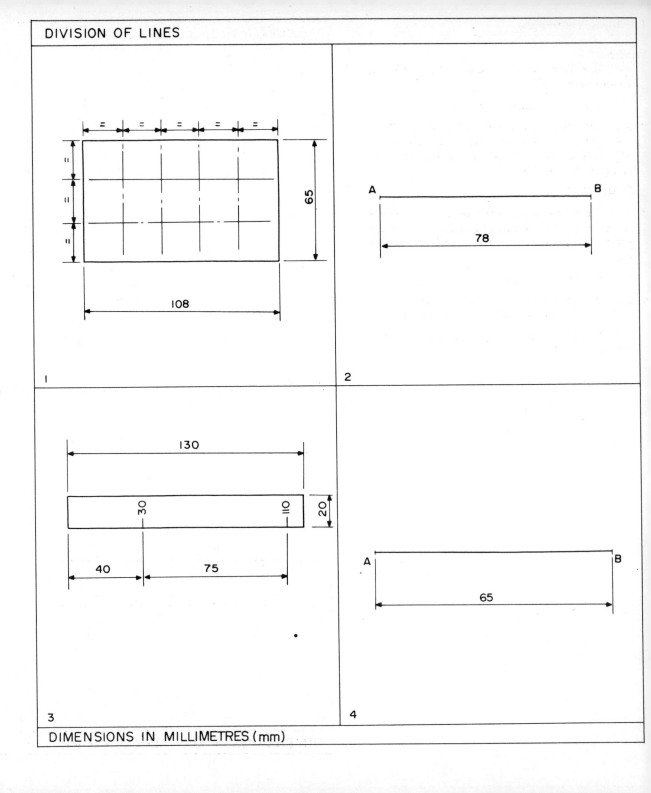

DIVISION OF LINES

DIMENSIONS IN MILLIMETRES (mm)

Tangents and circles

Panel No:

1. The panel illustrates in line form a see-saw. Using a scale of 1 mm to 10 mm draw the diagram showing clearly the construction for finding the point of contact at 0.

2. Round off this given angle by drawing a circle of 25 mm radius to touch both arms. SHOW YOUR CONSTRUCTION.

   E. Midland Reg. Ex. Board.

3. The figure shows the plan of a centre square, a tool used for marking diameters on the ends of cylinders. Draw TWICE FULLSIZE the given plan SHOWING CLEARLY YOUR CONSTRUCTION to determine the centres of the semi-circle and the 25 mm radius arc.

   E. Midland Reg. Ex. Board.

Ex.4 Describe THREE circles, each one touching the other two externally, their radii being 10 mm, 20 mm and 25 mm respectively.

   N. Western Sec. Schools Ex. Board.

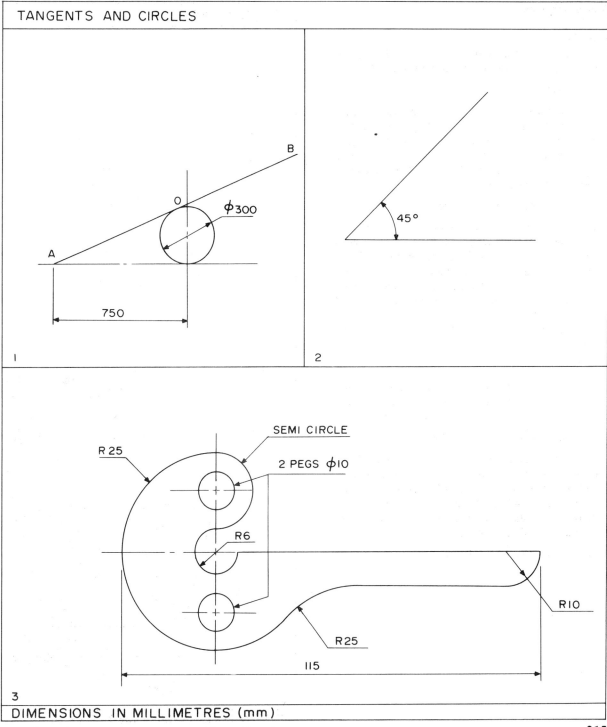

Regular polygons

Panel No:

1. AB in the position shown is one side of a regular hexagon. Construct the hexagon on side AB.

   W. Midlands Ex. Board.

2. You have a table with a square top and you wish to convert this to a more interesting shape, namely an octagon. Show how to mark out the largest possible regular octagon which can be cut from a square. Use a square to the given dimensions and show your construction.

   E. Midland Regional Ex. Board.

3. This panel shows in diagrammatic form part of a revolving feed mechanism. In operation the rollers engage with paper resting on the platform PT and feed it forward in five equal stages. The platform PT is tangential to the roller path. Draw, full size, the pentagon and the star with the five rollers and then draw the 150 mm long tangential platform from the given point T. All construction must be clearly shown.

   E. Anglian Ex. Board.

Ex.4 Construct a regular octagon on a base line 25 mm long and draw the inscribed circle. Measure and state the diameter of this circle in millimetres.

   N. Western Sec. Schools Ex. Board.

Ex.5 You are required to machine a metal bar of circular cross-section to one the cross-section of which will be in the shape of a regular pentagon with a side of 40 mm. Construct the pentagon and mark in and state the smallest diameter necessary for the original bar to contain it.

   Middlesex Regional Ex. Board.

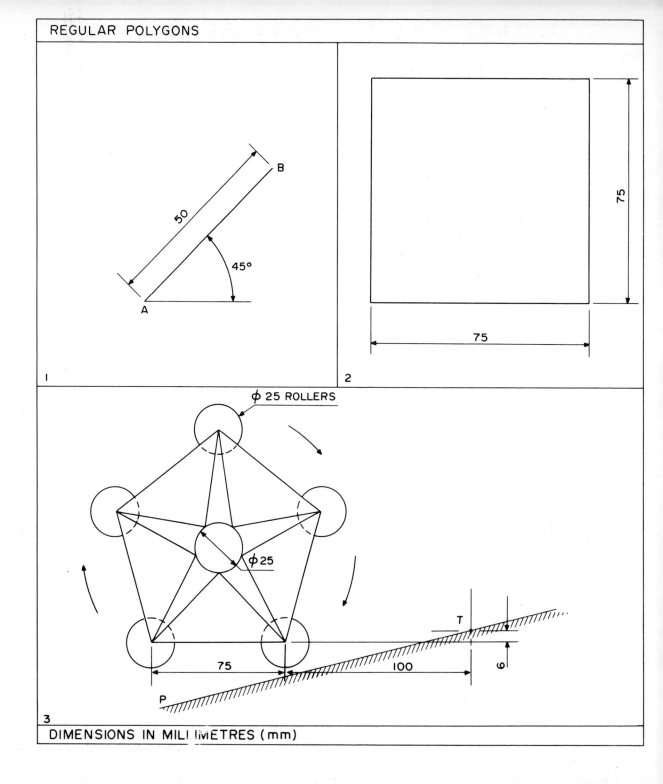

Equivalent area and similar figures

Panel No:

1. Convert the given triangle into a rectangle of equal area.

    E. Anglian Ex. Board.

2. The panel shows the plan of a flowerbed to be drawn using a scale of 1 m to 10 mm. Convert this to a square flowerbed of equal area clearly showing your construction. Measure and state the length of the side of the square in metres correct to one decimal place.

3. The panel shows a template to be drawn full size. The sides ED and DC are at right angles and of equal length. Show your construction for obtaining the point D. Draw a similar template with the base AB measuring 110 mm and the remaining sides in proportion.

4. The panel shows the section of a moulding. Determine geometrically the shape of a similar moulding whose height is 75 mm and width is 60 mm.

    N. Western Sec. Schools Ex. Board.

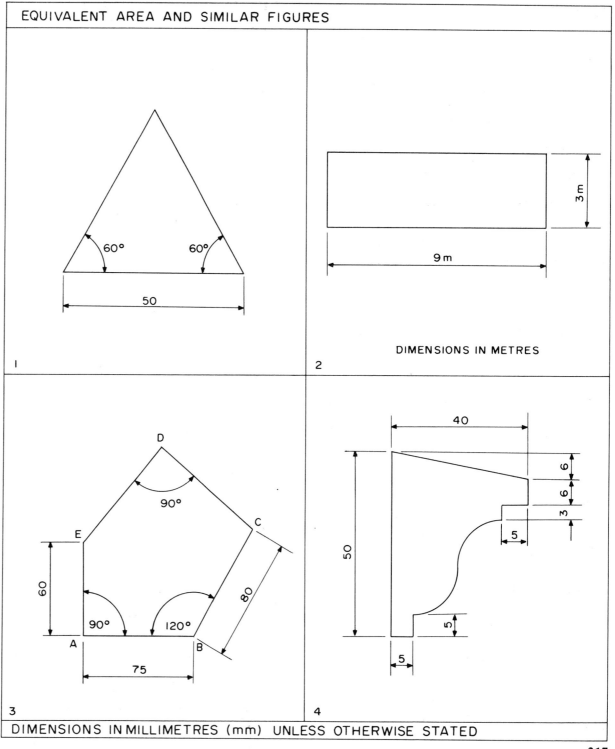

Loci

Panel No:

1. The panel shows a crank and piston. Draw full size the locus or path of point C for one revolution of OA. B slides in a horizontal direction along the line OX.

   OA = 40 mm; AB = 100 mm; AC = 55 mm

2. The drawing represents a wheel of a cycle with R a point on the tyre. Construct the locus of the point R as the wheel makes one revolution along the road. For the purpose of this exercise make the wheel diameter 60 mm. What name is given to this curve?

   N. Western Sec. Schools Ex. Board.

3. The figure shows a section through the hull of a boat. The curve follows a path in which its distance from the given point P is always equal to its vertical distance from the ground line XY. To a scale of 1 metre to 25 millimetres construct this section.

   W. Yorks & Lindsey Regional Ex. Board.

Ex.4 A dish is to be made from an elliptical piece of metal. The piece of metal from which the ellipse has to be cut is a rectangle 150 mm by 90 mm, and the ellipse has to be the largest possible from this piece.

   a) Construct the ellipse.

   b) Find the two points on the major axis which make a total distance of 150 mm from any point on the curve.

   b) Name the points found in (b).

   d) State the straight distance, in millimetres, the points are apart.

   Middlesex Regional Ex. Board.

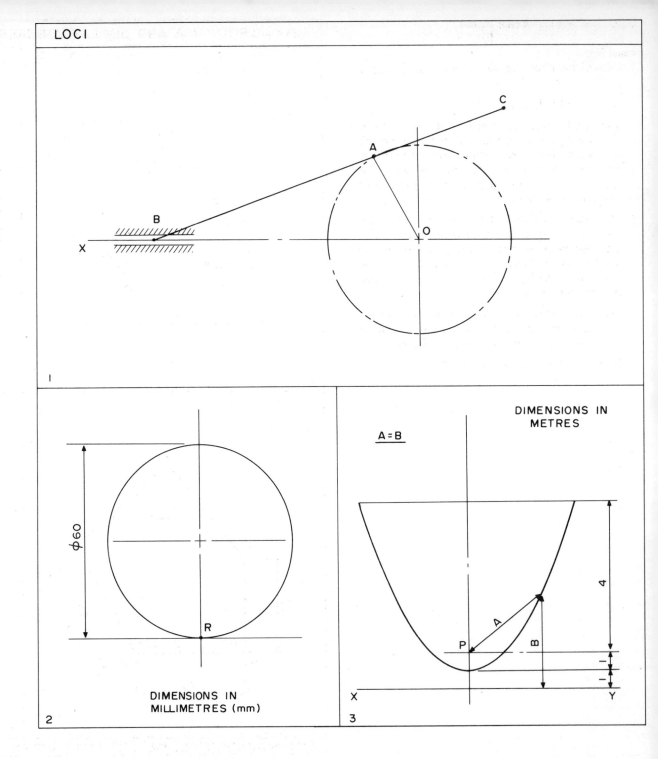

Interpenetration and developments

Panel No:

1. This panel shows a 600 mm square ventilating duct intersected by a 300 mm square duct. The axes of the ducts cross at 90°. Draw to a scale of 1 mm representing 30 mm the complete front elevation, showing clearly how you obtain the lines of the joint between the two ducts.

   Middlesex Regional Ex. Board.

2. The panel shows a lamp shade tilted at an angle of 30° to the horizontal. The cross section of the shade is a regular hexagon.

   Draw to a scale of ¼ FULL SIZE.

   a) The given elevation;

   b) A plan in projection with (a);

   c) The development of the sloping sides.

3. A small scoop is to be made to the dimensions given in the elevation. Draw the development of the shape of the metal required for the body of the scoop with the joint on AB. Ignore the thickness of the metal and do not allow for any overlap at the joint.

   Middlesex Regional Ex. Board.

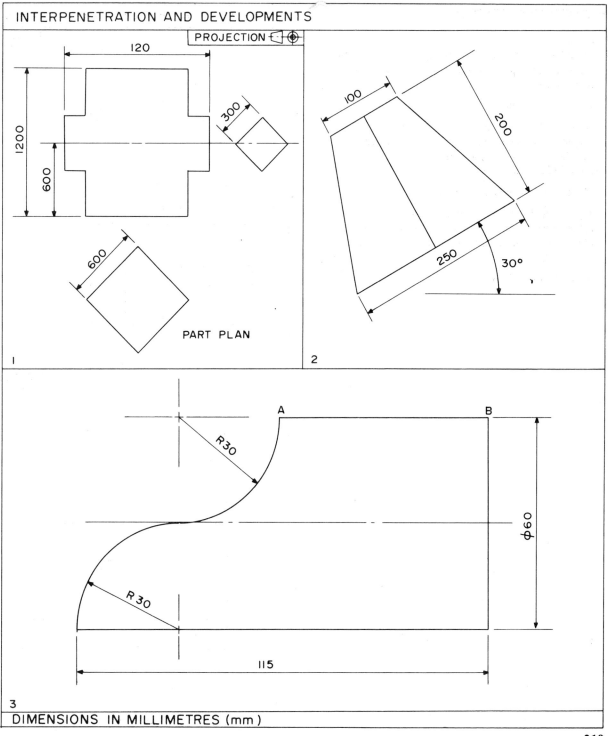

DIMENSIONS IN MILLIMETRES (mm)

Auxiliary views and true shapes

Panel No:

1. Complete on X_1Y_1 an auxiliary elevation of the metal packing piece shown in the two given complete views.

   E. Anglian Ex. Board.

2. A front elevation and plan of a half-lap dovetail projected in the 1st Angle is given. Draw an auxiliary elevation upon the new X_1Y_1 in the direction of the arrow.

   Middlesex Regional Ex. Board.

3. The panel shows the front view and end view of a cable cover. Draw full size the given views, the plan, and a true shape of section on the cutting plane S-S.

   E. Anglian Ex. Board.

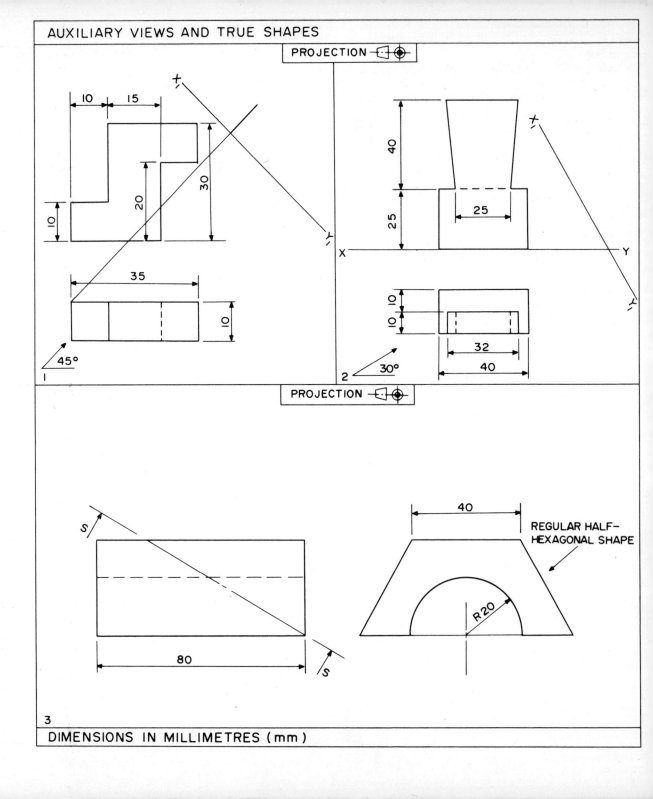

True length and true shapes

Panel No:

1 The figure represents the plan and elevation of a mast supported by three stay wires. Draw the views and SHOW YOUR CONSTRUCTION to determine the TRUE LENGTH of one wire.

E. Midland Regional Board.

2 The panel shows a line diagram of a half open gate. Using a scale of 1 mm to 10 mm find by projection the length of material required for the diagonal brace AB. Measure and write down this length.
NOTE
In the plan view the gate will be at 45° to the closed position. Do not allow for thickness of material or joints.

3 The figure shows two elevations of a regular hexagonal bar machined as shown. The front elevation being a view before the 12 mm slot is cut.

Part I
Draw full size showing all construction lines and hidden detail:

1 the given front elevation with the correction for the 12 mm slot;

2 the given end elevation;

3 a plan view in projection with and below 1.

Part II
Draw the TRUE shape of the sloping surface marked A.

Southern Regional Ex. Board (Northern).

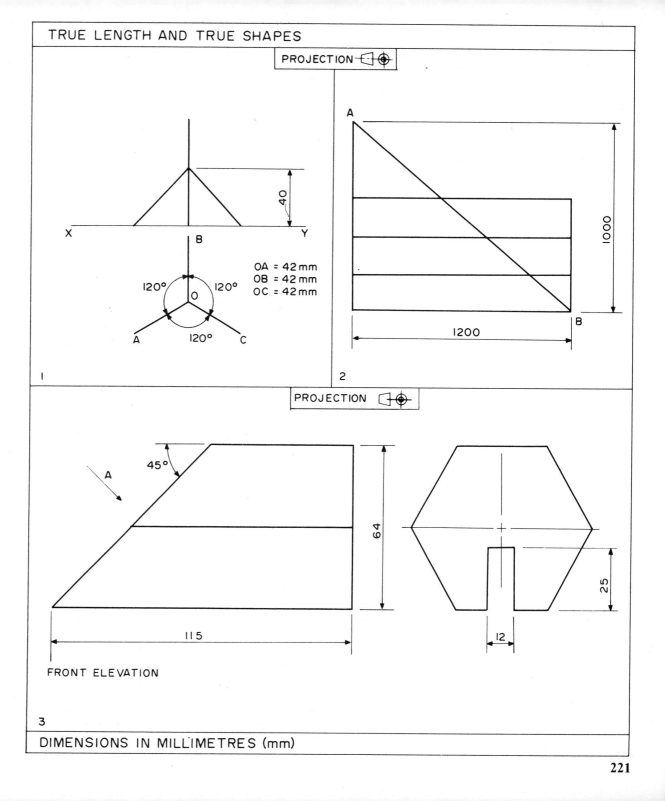

Index

	Page		Page		Page
Acute angle	70	Foci	108	Pitch	106
Altitude	14	Frustum	190	Pitch circle	54
Angles	2			Plain scale	24
Apex	14	Helix	106	Plan	114
Arc	54	Heptagon	38	Prism	136
Auxiliary views	126	Hexagon	30	Protractor	2
Axis	108	Hexagonal prism	184	Pyramid	134
		Horizontal plane	114		
Centre line	54	Hypotenuse	14	Quadrant	66
Chain line	54			Quadrilateral	6
Chord	54	Inscribed circle	92		
Circle	54	Inclined plane	134	Radius	54
Circumference	54	Internal tangent	88	Ratio	22
Circumscribing circle	92	Interpenetration	192	Regular polygon	30
Common tangents	88	Intersection	192	Representative fraction	24
Cone	188	Involute	100	Right angled triangle	14
Cross hatching	124	Isometric axes	136		
Cycloid	102	Isometric projection	136	Scalene triangle	14
Cylinder	106	Isosceles triangle	14	Scales	24
Cube	116			Sections	124
		Locus	94	Sector	54
				Side vertical plane	114
Developments	184	Major axis	108	Similar figures	48
Diagonals	143	Mechanism	94	Slant edge	208
Diagonal scale	26	Minor axis	108		
Diameter	54			Tangent	56
Division of lines	20	Oblique projection	178	Third Angle	120
		Obtuse angle	68	Trace	124
Elevation	114	Octagon	36	Triangles	14
Ellipse	108	Ordinates	172	True length	184
End elevation	114	Orthographic projection	114	Truncated	190
Equilateral triangle	14				
Equivalent area	42	Parabola	104	Vertex	14
Escribed circle	92	Parallel lines	20	Vertical plane	114
Exterior angles	3	Pentagon	40		
External tangents	88	Perpendicular bisector	8	XY lines	114